Digital Agriculture

Daniel Marçal de Queiroz
Domingos Sárvio M. Valente
Francisco de Assis de Carvalho Pinto
Aluízio Borém • John K. Schueller

Editors

Digital Agriculture

 Springer

Editors
Daniel Marçal de Queiroz 🆔
Department of Agricultural Engineering
Universidade Federal de Viçosa
Viçosa, MG, Brazil

Domingos Sárvio M. Valente 🆔
Department of Agricultural Engineering
Universidade Federal de Viçosa
Viçosa, MG, Brazil

Francisco de Assis de Carvalho Pinto
Federal Universidade of Viçosa
Viçosa, Brazil

Aluízio Borém 🆔
Federal University of Viçosa
Viçosa, Brazil

John K. Schueller 🆔
Mechanical and Aerospace Engineering
University of Florida
Gainesville, FL, USA

0th edition: © The editors 2021

ISBN 978-3-031-14535-3   ISBN 978-3-031-14533-9   (eBook)
https://doi.org/10.1007/978-3-031-14533-9

This Springer imprint is published by the registered company Springer Nature Switzerland AG
The registered company address is: Gewerbestrasse 11, 6330 Cham, Switzerland

Foreword

Most of the urban population has a romanticized and distorted view of rural businesses, formerly called farms. If asked, many of them will respond by describing their childhood vacations on their grandfather's farm, where almost everything the family needed at the table was produced: rice, beans, corn, vegetables, meat, milk, eggs, and fruits, among others. This image of farms is a thing of the past. Today, managed as companies, agricultural properties have evolved and are focused on technified food production, seeking economic sustainability.

For many industries, the Digital Age brought changes at the beginning of the twenty-first century, resulting from the high processing capacity of computers, smartphones, the geolocation system, robotization, and other advances in information technology. The Digital Era brings new ways of organizing production, and optimizing operations and logistics, in addition to offering better ways to meet consumer demands. This technological wave led to the modernization of information and communication processes, becoming the driving force that has already transformed the healthcare and drug, telecommunications, and automotive industries, as well as banking, commercial activities, among others. Agriculture is beginning its digital transformation. The adoption of automation, high-tech sensors, cloud computing, decision-making algorithms, and the Internet of Things is creating digital agriculture, in which collected data is used to increase the efficiency of the resources used by agriculture: land, water, labor, and inputs, which are finite. Much of the collection of this data is being carried out by sensors installed on agricultural machines, drones, satellites, and other platforms. These data, often available in real time, are used for both monitoring and decision-making in agriculture. Some are saying that data is the new oil of the twenty-first century, such is its importance in all productive sectors, and in agriculture, it is no different. This agricultural revolution introduces the concept of smart farms, with the interconnection of agricultural machinery, equipment, and sensors.

With digital agriculture, the producer can monitor all processes on their farm 24/7 on smartphone or tablet, optimizing decisions according to spatial and temporal variability, leading to a new revolution in food production.

This book, *Digital Agriculture*, provides an excellent opportunity for everyone involved in agriculture to stay updated with this new era, as it shows how technology is transforming the science and art of food production. What path is agriculture now taking? That is the focus of this book.

Universidade Federal de Vicosa (UFV), Brazil, is honored to have brought together a select group of professors and researchers in this book to present the state of the art in rural business management using digital agriculture technologies. I hope you can use it to help you build an agriculture capable of satisfying the world's food demand in an environmentally responsible and sustainable way. Good reading!

President of UFV   Demetrius David da Silva
Viçosa, Brazil

Contents

Chapter 1
The Agriculture Eras

Aluízio Borém and Lucas de Paula Corrêdo

During the last centuries, agriculture has been reengineered to attend the ever-growing population food demand. But currently, it is going through a transformation never seen before. This is the main scope of the book. In agriculture history, four eras or phases can be identified (Figs. 1.1, 1.2, 1.3, and 1.4). Rudimentary agriculture, which has been called Agriculture 1.0, dates back to its beginnings about 10,000 years ago, until about 1920, when it was based on physical strength—manual labor and animal traction. This agriculture demanded a lot of labor and, therefore, limited the size of the cultivated fields and was mainly focused on subsistence, generating few marketable surpluses. Its subsequent era refers to the beginning of more tech intense agriculture, also called Agriculture 2.0, corresponding to the period from 1920 to 1990—when the adoption of technological packages, such as machines, fertilizers and improved varieties, and other technologies became more widespread. Producing more food per unit area was the goal, using good agronomic management practices and new inputs, such as pesticides and chemical fertilizers. Part of this era was the Green Revolution, led by Norman E. Borlaug, with its semidwarf varieties. The success of agriculture in this period was so great, generating huge food surpluses, that society lost the sense of agriculture's essentiality for human life. Before the start of agriculture, man spent most of their days looking for food in the wild. It should also be pointed that the divisions among eras are somewhat arbitrarily and not well precise.

At the beginning of the last century, still in Agriculture 1.0, a farmer used to produce food for about four people. In 1960, during Agriculture 2.0, each farmer produced food for 26 people, and currently, in Agriculture 3.0, it feeds more than 150 people (Kirschenmann 2020). That is why today, many urban dwellers have no idea what takes place for the production of food that and for it to arrive at their table.

A. Borém (✉) · L. de Paula Corrêdo
Federal University of Vicosa, Viçosa, Brazil
e-mail: borem@ufv.br; lucas.corredo@ufv.br

© The Author(s), under exclusive license to Springer Nature
Switzerland AG 2022
D. Marçal de Queiroz et al. (eds.), *Digital Agriculture*,
https://doi.org/10.1007/978-3-031-14533-9_1

Fig. 1.1 Physical strength—men and animals were the driving force of agriculture until 1920: Agriculture 1.0. (Source: Picture belongs to the author)

Fig. 1.2 Mechanization started Agriculture 2.0. (Source: Picture belongs to the author)

For most of them, food is simply something constantly available on supermarket shelves. Imagine, for example, a long-term space travel. Agriculture is being reinvented by NASA to be part of planning the long space journeys not only to produce food but also to renew oxygen. For example, due to the simplicity of the process, it was possible, first, to produce ice cream in space before lettuce, as reported by Khodadad et al. (2020). Many challenges remain to be solved to enable the production of food in space, as shown in this study carried out at the International Space Station (ISS). Agriculture is essential to human life!

Subsequently, associated to the technological usage in the farmer's fields, a new phase arose due to genetic engineering, taking from the laboratories to the farms improved varieties never thought by the growers, which reduced the use of pesticides and facilitated crop management. This phase, called Agriculture 3.0, covers the period from 1990 to 2020 and continues to generate new and surprising varieties every year, advancing with the objective of maximizing the use of natural resources.

A new phase in agriculture is beginning, Agriculture 4.0, which includes, among many other changes, the increased use of biological products, the inclusion of new species still in domestication in food production, and digital transformation. This book is focused on the digital transformation of agriculture.

Fig. 1.3 Pesticides and improved varieties started Agriculture 3.0. (Source: Picture belongs to the author)

Fig. 1.4 Drone flying over crops and robot navigating over a crop canopy, collecting information on plant health and production components, autonomously, using RGB, LiDAR, and other sensors: Agriculture 4.0. (Source: Picture belongs to the author)

The Digital Era began at the beginning of the twenty-first century, characterized by the rapid change of the traditional industry with the adoption of information technology. This new era has opened up new ways of organizing production, optimizing operations and logistics, and offering better ways to meet consumer demands. This technological wave led to the modernization of information and communication processes, becoming the driving force that has already transformed many industries: telecommunications, capital goods, health, services, transportation, automotive, among others (Clay and Kitchen 2018).

The digital transformation is advancing at a fast pace, and certainly its application in agriculture is essential to sustainably meet the global food demand of nine billion people, in addition to feed for animals, in 2050. The adoption of automation and robotics; soil, plant, and climate sensors; the processing and storage of data in the clouds; AI; and connectivity are at the basis of this new agricultural revolution, called Agriculture 4.0, creating what many believe to be a New Green Revolution. Gigantic data volumes, information in terabytes, are collected for the most efficient and sustainable use of finite agricultural resources, such as land, water, labor, and inputs. Much of the collection of this data is being carried out by drones, satellites, and other platforms (Table 1.1, which presents a relative comparison of different

Table 1.1 Platforms for data collection in farms

| Criteria | Plataforms[a] | | | | | |
	Tractor	Drones	Airplanes	Stationary platform	Balloons	Orbital[b]
Proved in tests	+++++	++++	++++	+++	+++	+++
Data available quick	+++	+++	++	+++++	++++	++
Easy relocation	++++	++++	+++++	+	++	+
Load capacity	+++++	+	+++	+++++	++	+
Crop safety	+	++++	+++++	+++	+++	+++++
Initial cost	\$\$\$	\$\$	\$\$\$\$	\$\$	\$\$	\$\$\$\$\$
Operational cost	\$\$	\$\$\$	\$\$\$\$	\$\$	\$	\$
Coverage	++	+++	++++	+	+	+++++
Resolution	+++	++++	++++	+++++	++++	+
Legally regulated	+	++	+++++	+	+	+

[a]+ Less evident, +++++ more evident, \$ more economical, \$\$\$\$\$ less economical
[b]Remote satellite sensing

Table 1.2 Some applications and cost of different sensors in agriculture

Sensors[a]	Applications	Cost[b]
RGB (400–700 nm)	Morphological traits, maturity, plant density	\$
NIR (800–1.200 nm)	Plant chemical composition, hydric stress	\$\$
Multi–/hyper-spectral (350–2.5400 nm)	Plant health, weed identification	\$\$\$\$
Thermal (8–13 nm)	Biotic and abiotic stresses	\$\$\$\$\$
LiDAR (905–1.550 nm)	Topography mapping, plant height	\$\$\$\$\$
Fluorescent	Chlorophyll quantification and senescence	\$\$\$\$\$
Radar RS (0.3–100 cm)	Crop identification, soil moisture	\$\$\$\$

[a]*RGB* photographic cameras, *NIR* near infrared, *LiDAR* Light Detection and Ranging
[b]\$ more economical, \$\$\$\$\$ less economical

types of platforms). The first questions that the farmer who is adopting digital technologies on his property asks, in general, are as follows: What kind of platform to use? What is the best sensor? These platforms (Table 1.1) are on the market because each one is better suited to a particular situation. Some provide great coverage, although the images generated are of lower resolution, such as satellites, while others—such as drones—provide less coverage, but the images in general have better resolution. Thus, the best option for the farmer is to consult with a specialized technician.

In these platforms, a varied type of sensors has been used, according to the monitoring objectives (Table 1.2). Sensors that capture the visible light spectrum are used to assess the architecture, morphology, maturation, plant population, and secondary characteristics associated with yield and the identification of crop pests and diseases. LiDAR, short for "Light Detection and Ranging," is the acronym for radars that send fast laser pulses to the surface to be scanned. A sensor measures the amount of time it takes for each pulse to return, mapping the topography of the surface and creating 3D models of objects or the environment. This is a technology that detects the exact size and position of objects. It has been used to help autonomous vehicles to identify objects, obstacles, crop volumes, cultivation lines, and people. Technologies such as LiDAR, among other sensors, such as gyroscope, electronic compass, and accelerometers, can be merged with GNSS using, for example, Kalman filter to produce more robust, accurate and safe guidance systems, with great potential for remote agricultural monitoring. Thermal sensors can measure photosynthetic parameters, while scanners can evaluate root characteristics. Agriculture 4.0 is taking steps toward what promises to be a long and productive journey of continuous learning.

The challenges of agricultural management are large and complex, and the profitability of agriculture is becoming less and less, allowing no errors or waste. Thus, precise methods of geolocation and quality agronomic data—with precision and accuracy—are essential for the agronomist to make technical recommendations in an optimized and environmentally responsible manner. In this context, the geographic information system (GIS) was designed to capture, store, manipulate, analyze, manage, and present spatial or geographic data. GIS applications allow users to perform interactive searches, analyze spatial information, and edit data on maps of importance for agricultural management. In the GIS, field information is georeferenced, in a way that allows growers, for example, to stratify the field in different areas.

Variations in crops can be of a spatial and, or, temporal nature and may occur due to the wide spectrum and complexity of the factors that affect plant development. These discrepancies in crops reflect growth patterns of the crop, weeds, and other organisms that cohabit in the area, which is why they must be considered in agricultural management. For example, spatial-temporal variability affects agricultural yield, as it relates to changes throughout the year, which are associated with changes in solar radiation, temperature, rainfall, cloud cover, and the locality, since each area has its own fertility, soil texture, pressure from pests and diseases, and topography. Therefore, it is also important to analyze the variations in yield and the factors that control it, in time and space and for each spot of the field or crops, allowing a management of factors that affect agriculture be better carried out. Chapter 3 analyzes the spatial and temporal variability of agriculture.

The survey of pests, diseases, and weeds carried out in farms is through samples taken by technicians in zigzagging the fields but is giving way to the use of aerial images that assess the entire area, which allows the manager to establish priorities and specific recommendations for each crop spot. With these images, it is possible to create maps of pest and weed infestation, maps of fertility and soil texture by targeted sampling, maps of application of inputs, and maps of yield, among many other agriculture factors. At harvest, the grower can analyze yield by several aspects—variety, soil, field, and maturity date—and determine harvest priorities, as well as to diagnose yield fluctuations according to variations due to soil type, fertilization, plant population, pesticides used, and management adopted, among other factors, to better plan the next crop (Ping and Dobermann 2005). Chapter 6 discusses the use of these maps for better production management (Fig. 1.5).

Fig. 1.5 Different digital maps used in agronomic recommendations. (Source: Picture belongs to the author and to Strider LLC)

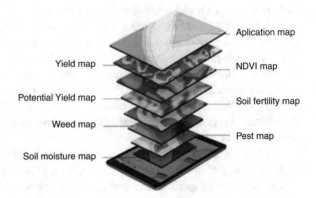

In the grain bin, sensors spread in the stored soybean remotely and in real time send humidity, temperature, and O_2 and CO_2 readings to panels that automatically adjust the aeration and relay this monitoring to the manager's smartphone. Chapter 7 addresses the use of different devices and sensors in agricultural management.

To manage this set of information, the agronomist needs to have access to it in a way that it can easily interpreted. Information not accurate or are temporarily outdated will result in technical recommendation that may be anachronistic and may not generate the expected corrective effect. For example, with digital agriculture, the grower can monitor the water layer applied to each slice of the pivot using a smartphone 24 × 7, using soil moisture sensors, which constantly send readings to different soil depths. Also, the pivot can be adjusted to apply the water layer at a variable rate, depending on the development of the crop, the unevenness of the soil, the variation in topography, etc. This theme will be described in detail in Chap. 8.

Automation and sensors have been used to improve efficiency and well-being in animal production. For example, automatic animal feed systems with growth and consumption control, which reduces waste and the need for the manager to be present on farms, are discussed in Chap. 9, which deals with digital animal science.

One of the greatest benefits of digital agriculture can be considered to support agronomists in defining their recommendations in real time, or based on data, precisely, and in the light of relevant information. With the pertinent data in a single panel and with interpretation algorithms, it becomes safer, faster, and more accurate to define the recommendations for each field spot. For example, the correct prescription of a certain insecticide depends not only on technical data of the crop and the infestation of the pests but also on the additional cost of application and the perspective of price of the agricultural product in the future market, the forecast of rains, and the proximity to the harvest, among other factors. Similarly, a fertilization plan must not only consider the data from the soil analysis as some farms today still do but also take into account other relevant variables and the possibility of applying it at a variable rate for each recommended nutrient and for each spot of the field. Thus, the role of digital agriculture is to support the agronomist in his/her diagnostics and recommendations and not to take his place.

In the hyperconnected world, where people, computers, and physical objects are interconnected to solve complex tasks, the concept of the Internet of Things, also called simply IoT, emerged. It is, therefore, the interconnection via the Internet of computing devices integrated in machines, instruments, sensors, equipment, and other devices, which allows these devices to receive and send data, in a process similar to a dialogue, mimicking a conversation between them. The IoT promises, in the coming years, to transform the sectors of education, health, finance, commerce, services, government institutions, and even agriculture. In the latter, the IoT will facilitate agronomic management, remote sensing (Maes and Steppe 2019), and the decision-making process, as it will allow growers to monitor, through their personal devices, remotely and in real time, the agricultural equipment, machines, and sensors, making them interact with each other. With this, the grower can, for example, continuously monitor irrigation, monitoring possible obstructions in the sprinklers, which are automatically triggered by signals emitted by soil moisture probes (Castrignano et al. 2020). In this way, growers can determine the specific requirements in real time for each location in the field, maximizing the efficient use of water, as well as providing growers with data for planning the use of agricultural inputs in a different way in time and space, in compliance with climate forecasts and crop development, which is also monitored by other sensors. IoT will make agriculture more efficient by: (i) reducing operating costs, (ii) increasing yield, and (iii) creating opportunities for the development of new products and services. It is estimated that, worldwide, there are already 50 billion interconnected devices, 24 billion of which belong to the IoT (Monsreal et al. 2019; CISCO Systems 2020). To facilitate understanding of how IoT works in agriculture, the infographic shown in Fig. 1.6 illustrates part of these applications. These and other IoT opportunities for agriculture are covered in Chap. 10.

However, what in fact is digital agriculture? Would it be the same as precision agriculture? For most specialists, the first is an evolution of the second and integrates digital technologies in the management of agricultural production processes. Thus, precision agriculture is a management strategy that gathers, processes, and analyzes temporal and spatial data and combines them with other information. All this in order to guide agricultural management, improving its efficiency, yield, quality, profitability, and sustainability, also reducing costs and risks, as well as enabling the management of large areas, remotely and in real time. Digital agriculture depends mainly on the implementation of data collection tools, which can accelerate and improve decisions in the adoption of good agronomic practices by the automating of various procedures for acquiring and, or, data processing (Heege 2013; Molin et al. 2015).

Some of the examples of digital transformation includes artificial intelligence (AI), machine learning, blockchain, big data, cloud computing, telemetry, augmented reality, deep learning, remote sensing, data mining, robotization, automation, and geographic information system (GIS), among other technologies. AI, an imitation of human intelligence, empowers machines, especially computer systems, with resources such as self-correction, learning, and reasoning. In agriculture, it will increase the precision and effectiveness in the maintenance and planting of crops, such as detecting pests, diseases, and weeds and deciding which agricultural

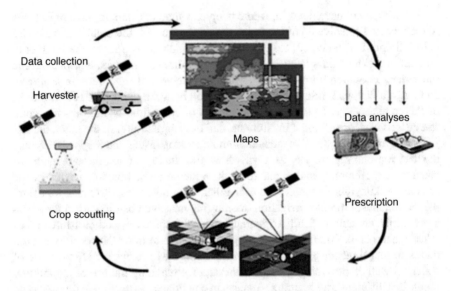

Data collection

Harvester

Maps

Data analyses

Prescription

Crop scoutting

Fig. 1.6 Infographic of the application of IoT in agriculture, with the intercommunication of different sensors and agricultural devices. (Source: L. Colizzi)

pesticides to apply in its control; avoiding the application of unnecessary pesticides and the development of resistance; and improving the harvest quality and precision. It also allows farmers to analyze a series of variables in real time, such as climatic conditions, water use, or soil conditions to support their decisions. Artificial intelligence emphasizes the development of machines that "learn" and work by simulating human behavior. It is widely used today as in such as facial and voice recognition, in autonomous vehicles, in the recommendation of items by the user's profile, in the autocorrection of texts in typing, in virtual assistants such as Alexa from Amazon, Siri from Apple, and Google's virtual assistant, who perform tasks such as calling people, sending messages, opening or closing a gate, turning lights on and off, searching on Google, and "chatting" with the user. This technology facilitates practically all activities of the grower.

At low sowing densities, modern soybean varieties exhibit compensatory mechanisms through increased branching, with a tendency to maintain yield. A shorter distance between rows and a higher sowing density have the benefits of improving the interception of solar radiation at the beginning of the cycle, reducing competition with weeds and, consequently, promoting increased yield. However, greater interlining and lower sowing density have the benefit of better disease control and reduced seed cost. Thus, the decision regarding the ideal sowing rate and row spacing is not simple, as it depends on the variety, severity of disease occurrence, weed infestation, and prevalence of days with sun high light intensity—no clouds, etc. How to make that decision? One of the applications of machine learning is to help minimize the number of variables without compromising the predictive models (algorithms), making the choice of the best combination of varieties and fertilization and interlining of each field easier for the manager.

AI techniques emerged after major advances in computer processing power and the accumulation of huge amounts of data. For example, there are already smartphones that can perform one trillion operations per second. These techniques have become essential for the technology industry, helping it to solve many problems arising from multivariate functions in real time. The intelligent use of data is one of the important characteristics of Agriculture 4.0, such as telemetry in the monitoring of tractors, sprayers, and harvesters, which allows the application of pesticides to be followed by the smartphone, tablets, or other devices. Chapters 11 and 12 describe cloud computing and various applications of AI in agriculture.

This fourth agricultural revolution introduces the concept of smart farms, with the interconnection of machinery, equipment, and sensors. It is agriculture evolving and allowing its managers to have tools that collect, store, and analyze data with agronomic relevance for practical and accurate recommendations. This is the scenario of global agriculture, in which time is scarce and having accurate, accurate, and real-time information in the palm of your hand is essential for the profitability of the business, which is why digital tools have become essential in the management of agribusiness. Platforms and panels are being improved so that there is compatibility of intercommunication between various agricultural equipment and these with the enormous amount of historical data from each area that need to be interpreted quickly and easily applicable in practice (Baio et al. 2018). With the use of digital tools, it is possible to monitor agriculture and to detect problems in real time, as the agricultural operation is carried out throughout the crop cycle. An example of this is planting with a variable rate of seeds, fertilizers, and application of pesticides and irrigation, controlled according to each spot of the field, or even allowing the monitoring of the speed of the harvest, the time remaining to complete the harvest, content of stored grain moisture, etc. (Molin et al. 2015). All this information can be obtained remotely, allowing the monitoring of field operations even when the manager is out of the property. Chapter 13 describes the use of these platforms in production management.

The management of farms is undergoing major modernization through automation, remote monitoring, augmented reality, and the use of accumulated data. To organize, process, analyze, and interpret this gigantic volume of available data, it has been created the term big data. Thus, this term deals with ways of using very large or complex data sets, so that the user can analyze them and generate relevant information for decision-making. Big data is related to five aspects: volume, variety, veracity, value and velocity, allowing greater resolution power and confidence of the agronomist in his/her decisions. Big data challenges include capturing, storing, analyzing, interpreting, sharing, transferring, securing, and visualizing these large data sets, as well as, of course, adapting them to regional particularities, climate change, new cultivars, etc. These data sets grow exponentially with the development of increasingly accessible and powerful devices and sensors (MAPA 2020; Viscarra Rossel and Lobsey 2016). The manipulation of these massive data sets often exceeds the capacity of a normal computers (hardware) or, even, software, creating enormous challenges. It should be noted that agriculture is a complex system and difficult to holistic analysis, as it has numerous interrelated variables, such as soil fertility; climate; varieties; incidence; and severity of pests, diseases, and weeds; agronomic practices, among other variables of the yield. It must also be recognized that agriculture is characterized by the incidence of risks and uncertainties. Each farm and field are different, each agricultural year is different, and the same equipment works in different fields and generates different results.

The data currently available for agricultural management include choice of variety, planting time, sowing density, soil fertility, and climatic conditions, as well as how much, when, where, and what inputs to use, among other factors. So, how to manage this diverse and complex system? For example, the choice of variety for a field depends on the sowing density and time, the type of soil and its fertility, the climatic forecast, and many other information and historical records of the area. When one of these factors is changed, the choice of the variety to be planted may fall on another one, which in turn requires different sowing density, thus implying the recommendation of different fertilization or specific agronomic management, etc. And how to make the right decision in real time in the face of so many variables (big data)? This only possible with the digital technologies! Analyzing, interpreting, and establishing specific recommendations and monitoring their implementation are one of the many benefits of the digital age for agriculture. Chapter 14 deals with the collection, systematization, sharing, and security of big data.

Digital agriculture also poses new challenges for agricultural companies and growers, which, over time, are being overcome. Some of these problems include training of the farm's human resources, technological updating, little connectivity in some rural areas, internet with slow connection, sensors with small coverage radius and limited spectrum of monitored characteristics, higher frequency and speed in the capture, and automation of the interpretation of the data (Bramley 2009). Chapter 15 presents a case study of the adoption of digital agriculture, addressing the challenges and opportunities in its implementation in Brazil.

Contrary to what some claim that the digital age would end job opportunities in the field, most experts understand that in a truly digital farm, similar to a factory that adopts robots, the use of labor is that will become more specialized. The repetitive services previously performed by man are left to automated systems, giving space for the best use of the human intellect. Freely, the human mind can dedicate itself to decision-making processes, like choosing the best agronomic practices, with the support of digital tools. In agriculture, no algorithm or AI will replace the practice and common sense of the field agronomist.

To support this agricultural revolution, many startups, called Agtechs, have been created and accelerated by investment funds and large multinationals, envisioning all the importance and economic potential of this business. These Agtechs have focused on areas such as sensors, digital agronomic consulting, drones, robotics, automation, processing, and big data analysis. From the point of view of professional career, people trained in this area are already in great demand in the market. Currently, it is estimated that annually 2.1 trillion dollars is spent on digital transformation technologies and services—double that of was invested in 2016 (Zastupov 2019).

In detail, but with accessible language, the following chapters discuss the main topics of digital agriculture. Through its implementation, a new "green revolution" will take place in agriculture. Be you, reader, also part of this transformation and contribute to a world with a greater abundance of food and less damage to the environment. The digital age is changing the practice of agriculture, but not its essence. Use it for efficient and responsible food production!

Abbreviations/Definitions

- AI: Artificial intelligence is the intelligence demonstrated by machines, as opposed to natural intelligence displayed by animals including humans. Some popular accounts use the term "artificial intelligence" to describe machines that mimic "cognitive" functions that humans associate with the human mind, such as "learning" and "problem-solving."
- Genetic engineering: Also called genetic modification, it is the direct manipulation of an organism's genes using biotechnology. It is a set of technologies used to change the genetic makeup of cells, including the transfer of genes within and across species boundaries to produce improved or novel organisms.
- Geographic information system (GIS): It is a type of database containing geographic data (i.e., descriptions of phenomena for which location is relevant), combined with software tools for managing, analyzing, and visualizing those data.
- Green Revolution: It was a period that began in the 1960s during, which agriculture in India was converted into a modern industrial system by the adoption of technology, such as the use of semidwarf high yielding varieties. It is led by agricultural scientist by Norman E. Borlaug, which leveraged agricultural research and technology to increase agricultural productivity in the developing world.
- Internet of Things (IoT): It describes physical objects (or groups of such objects) that are embedded with sensors, processing ability, software, and other technologies that connect and exchange data with other devices and systems over the Internet or other communications networks.
- Light Detection and Ranging (LiDAR): It is a device that detects the position or motion of objects and operates similarly to a radar but uses laser radiation rather than microwaves.
- Telemetry: It is the in situ collection of measurements or other data at remote points and their automatic transmission to receiving equipment (telecommunication) for monitoring.
- Terabytes: It is approximately a thousand billion bytes. A unit of computer information consisting of 1,000,000,000,000 bytes.
- Visible light spectrum: It is the portion of the electromagnetic spectrum that is visible to the human eye. Electromagnetic radiation in this range of wavelengths is called visible light or simply light. A typical human eye will respond to wavelengths from about 380 to about 750 nanometers.

Take Home Message/Key Points

- Agriculture was developed about 10–12,000 years ago and evolved slowly in the next few millenniums.
- Agriculture has evolved fast in the last centuries.
- Four eras can be identified in agriculture history.
- Digital transformation is currently permeating all aspects of agriculture to produce more sustainably and to allow better farming management.

References

Baio FHR, Molin JP, Povh FP (2018) Agricultura de precisão na adubação de grandes culturas. In: Nutrição e adubação de grandes culturas, 1st edn. Unesp, Jaboticabal, pp 351–379

Bramley RGV (2009) Lessons from nearly 20 years of Precision Agriculture research, development, and adoption as a guide to its appropriate application. Crop Pasture Sci 60:197–121

Castrignano A, Buttafuoco G, Khosla R, Mouazen, Moshou D, Naud O (2020) Agricultural internet of things and decision support for precision smart farming. Elsevier, New York, 470p

CISCO Systems (2020) IoT case studies. Available at https://www.google.com/aclk?sa=l & ai=DChcSEwiimLvT_pTnAhVG1sAKHUKDC2sYABABGgJpbQ & sig=AOD64_2rPf673LQrAu2c_ZrpkpcLgfy0Gg & adurl= & q= & ved=2ahUKEwiNrbLT_pTnAhWKbc0KHcWvBhoQqyQoAHoECBMQBw. Acessed 25 abr 2020

Clay D, Kitchen N (2018) Precision agriculture basics. American Society of Agronomy, New York, 287p

Heege HJ (2013) Precision in crop farming. Springer, New York, 363p

Khodadad CLM et al (2020) Microbiological and nutritional analysis of lettuce crops grown on the international space station. Front Plant Sci. Available at https://doi.org/10.3389/fpls.2020.00199

Kirschenmann, F. (2020) How many farmers we will need? Iowa State University, Iowa. Available at https://www.extension.iastate.edu/agdm/articles/others/KirJan01.htm

Maes WH, Steppe K (2019) Perspectives for remote sensing with unmanned aerial vehicles in precision agriculture. Trends Plant Sci 24:152–164

MAPA (2020) Agricultura digital e de precisão. Available at http://www.agricultura.gov.br/assuntos/sustentabilidade/tecnologia-agropecuaria/agricultura-de-precisao-1/agricultura-de-precisao

Molin JP, Amaral LR, Colaço A (2015) Agricultura de precisão. Oficina de Textos, São Paulo, 238p

Monsreal M, Calatayud A, Villa JC, Graham M, Metsker-Galarza M (2019) Estimating the benefits of the internet of things for supply chain performance and management—a conceptual framework. In: Transportation Research Board 98th annual meeting. Proceedings … vol 6, pp 102–27

Ping JL, Dobermann A (2005) Processing of yield map data. Precis Agric 6:193–212

Viscarra Rossel RA, Lobsey C (2016) Scoping review of proximal soil sensors for grain growing. CSIRO, 216p

Zastupov AV (2019) Innovation activities of enterprises of the industrial sector in the conditions of economy digitalization. In: Ashmarina S, Mesquita A, Vochozka M (eds) Digital transformation of the economy: challenges, trends and new opportunities. Springer Press, New York, 462p. Available at https://doi.org/10.1007/978-3-030-11367-4

Chapter 2
Global Navigation Satellite Systems

**Daniel Marçal de Queiroz, Domingos Sárvio M. Valente,
and Andre Luiz de Freitas Coelho**

2.1 Introduction

Accurately knowing the location where we are on the Earth's surface has always been a desire of humanity. This problem became even more important with the development of maritime ocean navigation centuries ago. In the twentieth century, military forces looked for the development of positioning systems for ships, submarines, and airplanes and developed systems based on land-based stations that transmitted signals at certain frequencies. From the launch of the first artificial satellite, which occurred in 1957, the opportunity for the development of satellite positioning system was opened. Late in the twentieth century, two satellite positioning systems, the GPS (Global Positioning System) implemented by the US Department of Defense and the Global Orbiting Navigation System (GLONASS) implemented by Russia, became fully available for civilian use. At the beginning, both systems had military purposes, but the systems were gradually opened to civil applications.

Although there are multiple systems, the term GPS is often generally used to designate any satellite positioning system because the American system was the first to become fully operational, widely publicized, and available. As other satellite positioning systems appeared, we started to use the term GNSS (Global Navigation Satellite Systems). Table 2.1 shows GNSS systems implemented or in implementation. It is important to mention that the number operational satellites are constantly changing due to new launches and retirements.

The Indian positioning system NavIC (Navigation with Indian Constellation) and the Japanese QZSS (Quasi-Zenith Satellite System) are regional positioning

D. Marçal de Queiroz (✉) · D. S. M. Valente · A. L. de Freitas Coelho
Universidade Federal de Viçosa, Viçosa, Brazil
e-mail: queiroz@ufv.br; valente@ufv.br; andre.coelho@ufv.br

© The Author(s), under exclusive license to Springer Nature
Switzerland AG 2022
D. Marçal de Queiroz et al. (eds.), *Digital Agriculture*,
https://doi.org/10.1007/978-3-031-14533-9_2

Table 2.1 Global satellite navigation systems available in 2021

Name	Description	Owner	Number of satellites in operation currently	Date of declaration of fully operational
GPS	Global Positioning System	USA	30	04/27/1995
GLONASS	Global Orbiting Navigation Satellite System	Russia	24	1995
Galileo	Galileo Navigation System	European Union	22	2020
BeiDou	BeiDou Navigation Satellite System	China	44	2020
NavIC	Navigation with Indian Constellation	India	8	2018
QZSS	Quasi-Zenith Satellite System	Japan	4	2024

systems and provide positioning in the country of origin and in its surroundings. QZSS, which currently has four satellites in orbit, will have seven satellites by 2024.

The European system Galileo is a global positioning system characterized by being designed for civilian use, so it has resources focused on this type of use. It was created by the European Space Agency. This system intends to give 1 m horizontal and vertical precision.

The development of satellite positioning systems has played an important role in the development of precision agriculture and now in digital agriculture. That is because in precision agriculture, for most operations, we need to know continuously where the machine is. Early work in precision agriculture before the availability of reliable and accurate GNSS used other systems, which were limited in accuracy and involved substantial costs. Satellite positioning systems have made it possible to locate agricultural machinery 24 h a day at a reasonable cost.

Digital agriculture is heavily dependent on satellite positioning. Digital agriculture platforms inform those responsible for the operational management of the machines, the positioning of each machine in the field. If there is connectivity in the field, the digital agricultural platforms provide real-time machine positioning. The data generated by the machines are georeferenced through positioning systems. Therefore, there is no digital agriculture without satellite positioning system.

In this chapter, we will show how satellite positioning systems work focusing on the particularities that the professional working in digital agriculture needs to understand. However, before we describe how these systems work, we will address the issue of coordinate systems that are used to define the position of a point, as this is important for us to understand how positioning systems work.

2.2   Projection and Coordinate Systems in Positioning

To obtain the position of a point in space, that is, in three dimensions, we need three coordinates. These coordinates can be expressed in the geographic coordinate system or using a projection system. In the geographic coordinate system, the position of a point is defined by two angles, latitude and longitude, and by the elevation relative to a reference surface. When a projection system is used, the points are projected onto a surface, and then that surface is transformed into a X,Y plane, while the point elevation represents the third coordinate.

The coordinates of a point are not enough to define the positioning, as it is necessary to define which reference system is used to obtain the coordinates. There are hundreds of reference systems being used in the world. The main reference systems currently used are geocentric, meaning that the origin of the system is at the Earth's center of mass. Examples of this type of systems are WGS84 (WGS is an acronym meaning World Geodetic System) and SIRGAS2000 (*Sistema de Referência Geocêntrico para as Américas* or Geocentric Reference System for the Americas).

In addition to defining the origin, reference systems define the parameters related to the ellipsoid model. This model is used to represent the surface that best approximates the surface that has gravitational potential equal to the average sea level. This surface, which is irregularly shaped, is called geoid. The ellipsoid is defined by its semi-major axis and flattening. Flattening is a dimensionless parameter that is obtained by dividing the difference between the semi-major axis and the semi-minor axis by the semi-major axis. Table 2.2 presents examples of reference systems used to express the coordinates of a point in South America.

In the projection system, one of the most common ways to represent the position of a point is through UTM (Universal Transverse Mercator) coordinates. In this system, the planet Earth is divided into 60 longitudinal projection zones with 6 degree longitude each that is numbered from 1 to 60 and in 20 latitudinal zones from latitudes 80° S to 84° N that are named by letters C to X. Figure 2.1 shows the longitudinal UTM zones to different parts of the world.

In the UTM coordinate system, for each latitudinal zone, the points on the Earth's surface are projected on a cylindrical surface, and that surface is converted to a flat

Table 2.2 Parameters of different reference systems used to define the coordinates

Reference system	Reference ellipsoid	Major semi-axis (m)	Inverse of flattening	Reference meridian	EPSG[a]
WGS84	WGS84	6378137.0	298.257223563	IERS[b]	4326
GRS80	GRS80	6378137.0	298.257222100882711	IERS	
SIRGAS2000	GRS80[c]	6378137.0	298.257222101	IERS	4674
SAD69	GRS67	6378160.0	298.25	Greenwich	4291

Source: Van Sickle (2020)

[a]Code that defines the geodesic parameter database of EPSG (European Petroleum Survey Group)
[b]Reference Meridian of IERS (International Earth Rotation and Reference Systems Service)
[c]GRS (Geodetic Reference System)

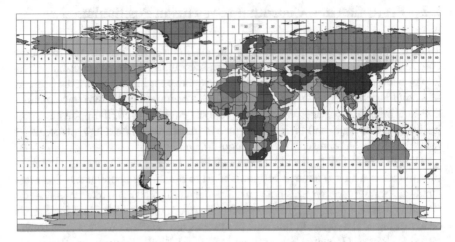

Fig. 2.1 UTM zones illustrating the position and the numbering of the longitudinal zones

surface. UTM coordinates are repeated for each zone; therefore, when using the UTM coordinate system, it is necessary to refer, in addition to the reference, to which zone the coordinate is being presented. As an example, the UTM projection system for the city of Viçosa, MG, could be referred to as WGS84 UTM 23S. In this example, the WGS84 reference system and the UTM projection system in zone 23 of the southern hemisphere (S) were specified.

The point where the central meridian line of each zone crosses the equator line, as shown in Fig. 2.2, has X-coordinate of 500,000 m. The X-coordinate decreases to the left of that point and increases to its right. The points in the northern hemisphere have Y-coordinate 0 m N at the equator, and this coordinate increases from the equator to the north pole. Points in the southern hemisphere has a Y-coordinate of 10,000,000 m S at the equator, and this coordinate decreases as it approaches the south pole. One advantage of using the UTM projection system is the metric unit of the coordinates, which makes it easier to interpret the data and define interpolation parameters for maps.

2.3 How Satellite Positioning Systems Work

Global navigation satellite systems are formed by three segments: spatial, control, and user. The spatial segment comprises the constellation of satellites that make up the given system and transmit signals to the other segments. The control segment consists of stations located at strategic points of the Earth that controls the satellites. The user segment is composed of the receiving device that receives and decodes the signal from the satellites and calculates the position and time. The GNSS systems listed in Table 2.1 have some differences between them. But many of the concepts

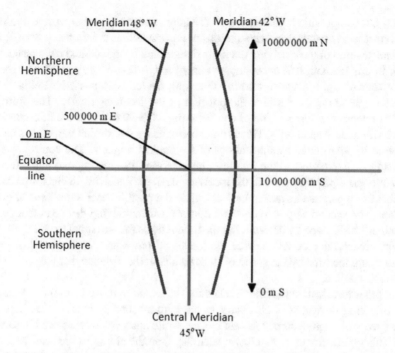

Meridian 48° W

Meridian 42° W

10 000 000 m N

Northern
Hemisphere

500 000 m E

0 m E

Equator
line

10 000 000 m S

Southern
Hemisphere

0 m S

Central Meridian
45°W

Fig. 2.2 Layout of the UTM coordinate system taking Zone 23 as an example

are the same. GPS will be used to provide examples below, but the other systems tend to have somewhat similar characteristics.

Each GNSS positions its satellites forming a constellation with a certain arrangement. Satellites are distributed in orbits, and the number of orbits can also change from system to system. Global positioning systems are designed to ensure that, at any time and in any position on Earth, there is always a minimum number of satellites to calculate the position. For example, GPS is designed to have a constellation with 24 satellites, distributed in six orbits, with four satellites in each orbit. These orbits have a tilt angle of 55° relative to the plane defined by the equator. GPS satellites move at an altitude of 20,200 km, which causes them to perform a turn around the earth in approximately 11 h and 58 min.

Control segment stations have the functions of monitoring satellites, adjusting their orbit and adjusting the atomic clock of satellites and sending them parameters that will be retransmitted to users. The communication role between these stations and satellites is carried out by a single station of the system, called the master station. This station centralizes all the control work of the positioning system.

The user segment is composed of the receivers that receive the signals from the satellites through their antennas. Each receiver has a processor to decode the signals and to calculate the position of the receiver. There are many different types of receivers with different signal receiving and processing capabilities. Those receivers, which can receive and process signals from multiple types of satellites, are more costly but can provide higher accuracies. Many receivers can also receive and process correction signals from base stations or other satellites to improve the accuracy.

The position of the receiving device is determined by the method called trilateration. In this method, it is necessary to know the position of the satellites in space, then determine the distance that the receiving device is from each satellite, and, based on these data, determine the position of the receiving device. Therefore, in order to determine the position of a point using a GNSS receiver, the first step is to know where each satellite is. This is done based on a set of orbital parameters called ephemeris, determined by the stations of the control segment. The control system sends these parameters to the satellites, and the satellites incorporate this information into the signal they send to the receiving devices. Receiving devices are able to extract these parameters and to then calculate the position each satellite is at every instant. The second step is to determine how far the receiving device is from each satellite. This is done by determining the time it takes for the signal to exit the satellite and reach the receiving device. As the signal propagates at the speed of light, multiplying the time by the speed of light results in the distance that the receiver is from the satellite.

In order to calculate the time it takes for the signal to move from the satellite to the receiving device, it is necessary to have high-accuracy clocks. Satellites are equipped with high-accuracy atomic clocks, while quartz clocks are used in receivers. This quartz clocks have a lower accuracy than the clocks of the satellites. The lower accuracy, even if it is equal to thousandths of seconds, generates high errors. Remember, this number will be multiplied by the speed of light to determine the distance.

If the time determined by the receiving device had the same accuracy as the time determined by the satellite clock, it would be sufficient to receive the signal from three satellites to determine the position of the point (x, y, z) at which the receiving device is. To correct the error of the receiver clock, we need to receive the signal from a fourth satellite. Therefore, in satellite positioning systems, it is necessary to receive signals from at least four satellites to accurately determine the position (x, y, z) and time (t). Receiving signals from more satellites allows statistical processing in the receiver to further increase accuracy.

Each GNSS system transmits the signals at specific frequencies. For example, GPS is designed to work with two frequencies, L1 of 1575.42 MHz and L2 of 1227.60 MHz. With the modernization of this system, satellites will transmit signal at four different frequencies. A signal at the L2C frequency, which is an L2 frequency for civilian use, and a signal at the L5 frequency (1176.45 MHz) will be made available. To facilitate the understanding, we will explain how the system works from the L1 and L2 frequencies.

Most receiving devices only pick up the L1 frequency signal. At the L1 frequency, two signals are modulated, the C/A (course acquisition) code and the P (Precision) code. All receivers pick up C/A code, while P code is restricted to authorized users. The C/A code is generated at a rate of 1.023 million bits per second (1.023 MHz). Each satellite has a specific code that has 1023 bits. This code is repeated every millisecond. The P code, also unique for each satellite, is generated

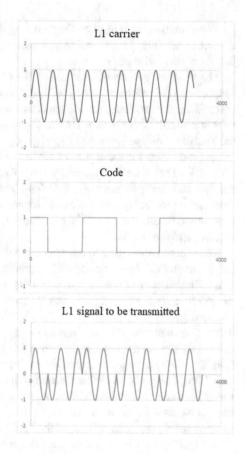

Fig. 2.3 Scheme showing how the signal is generated by the phase change process

at a rate of 10.23 million bits per second (10.23 MHz) and is repeated every 259 days (Grewal et al. 2020). These codes are generated in binary format, that is, they are composed of 0 and 1. C/A and P codes are incorporated into the L1 wave by a process called phase modulation. With each change from 0 to 1 or from 1 to 0, a phase change of 180° is applied on the L1 wave. Figure 2.3 shows a diagram illustrating how the L1 signal is generated. Together with the C/A code, each satellite sends the data that make it possible to estimate its orbit (ephemeris), the orbit data of all satellites, called almanac, correction for signal delay when crossing the ionosphere and correction to the satellite clock.

In the L2 band, only the P code is transmitted. The advantage of receiving this type of signal lies in the fact that the signal delay when crossing the layers of the atmosphere depends on the frequency squared and, with the reception of two frequencies, it is possible to estimate this delay and reduce the positioning error.

All satellites of the GPS system emit signals at the same frequency, and in the message sent by the satellite, the receiving device can distinguish from which satellite the received signal has come. In the GLONASS system, the signal frequency of each satellite is different, and all satellites emit the same signal (Grewal et al. 2020).

There are two ways used by the receiving devices to determine the distance from it to each satellite. One is called processing based on C/A code, and the other is based on the carrier phase.

The simplest method of determining the time of signal displacement between the satellite and the receiver is based on the C/A code. In this method, called code-based processing, the receiving device generates the C/A signal that each satellite transmits and simultaneously receives the signal sent by the satellite. The signal sent by the satellite is delayed from the signal generated by the receiving device. This delay is the time taken for the signal to move from the satellite to the receiving device (Fig. 2.4). This time multiplied by the speed of light gives the distance that the receiver antenna is from the satellite. The problem of this type of processing lies in determining the delay suffered by the signal, as an error of one microsecond in this time would mean an error of 300 m in the determination of the distance between the receiver and the satellite.

The other method used to determine the distance between the receiver and satellite is the method based on the carrier phase. This method involves the determination of the number of whole waves and the wave fraction between the receiver and the satellite. The wavelength of the L1 band (frequency of 1575.42 MHz) can be obtained by dividing the speed of light by the frequency of the wave. This means that the wavelength of the L1 band is 19 cm (Eq. 2.1). As this method makes it possible to determine fractions of the wave, this means that the error in determining the distance between the receiver and the satellite with current technologies can be less than 1 cm. The problem of this method is the determination of the number of whole waves between the receiver and the satellite, a process called ambiguity. For this to be possible, it is necessary to have a second receiving device fixed in a location with known coordinates, which is called the base receiver. This second receiver is used to obtain ambiguity for each satellite. The receiving device used to determine the

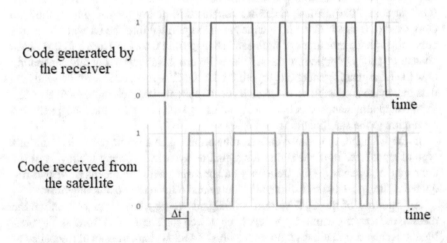

Fig. 2.4 Scheme showing the time interval the signal takes between the satellite and the receiving device when using the code-based positioning method

positioning of the points, called a rover, must be close to the base, usually at a distance of less than 10 km from it, the accuracy is high. If the base is less than 10 km from the rover and if there is a radio link between the base and the rover to communicate the location data, the base calculates to the rover, such that the positioning is done instantly, and the system is called Real-Time Kinematics (RKT). RTK correction systems are widely used in modern agriculture. In RTK-type systems, the error is typically less than 5 cm.

$$\lambda = \frac{c}{f} = \frac{300\,10^6}{1575.42\,10^6} = 0.19\,m \tag{2.1}$$

where:

λ = Wavelength, m;
c = Speed of light, m/s;
f = Wave frequency, Hz.

2.4   Satellite Positioning Errors

In satellite positioning systems, there are basically two types of errors. The first is the one caused by the user and the other is intrinsic to the system.

The error sources associated with the user are basically related to the setting of the receiving device. The most sophisticated receivers depend on a number of settings that can affect the positioning error, for example:

(a) Choice of the reference system to be used. The possibility of using different reference systems for mapping the same area can lead to errors when transferring data from the receiver to the system that will be used to process these data.
(b) Choice of an elevation mask to capture the satellite signal, thus avoiding the use of signals with too many errors (satellites with elevation angle below the mask).
(c) Mission planning (moment of collection of georeferenced points) based on the location and time data of each satellite (system almanac) makes it possible to work at times that are more favorable to the use of GNSS, if the almanac is not very old.

These errors can be reduced with proper user training. However, there are sources of errors intrinsic to the satellite positioning system, which are the multipath errors, the ones due to the ionosphere and troposphere, the ones due to the satellite's orbit, and the ones due to the satellite clock.

The multipath error occurs when the signal is reflected by objects in the path from the satellite to the receiver, not directly reaching the receiver, thus increasing the calculated distance between the satellite and the receiver. More sophisticated receivers and antennas can discriminate the multipath signal and use only the signal coming directly from the satellite. There should always be a greater concern with the accuracy of positioning when working near buildings that may reflect the signal. This problem is generally not a common situation in agriculture. The approximate multipath error is around 1 m.

The error due to the ionosphere occurs because of the delay or deviation that the satellite signal suffers from the ions and electrons present in the atmospheric layer located between 50 and 1000 km above the Earth's surface (ionosphere). As the ionosphere is not homogeneous and varies with the time of year and time of day, the signal may be influenced more in certain regions than in others, more in certain periods of the year than in others and more at certain times of the day than at others. Monitoring and control stations calculate correction factors that are transmitted to correct the effect of the ionosphere. However, not all receiving devices use algorithms that make use of this information. The delay error in the ionosphere can be approximately 5 m if it is not handled properly by the system.

The troposphere, the tropopause, and the stratosphere are three layers that make up the layers of the atmosphere that go up to 50 km above the Earth's surface. These layers delay the signal emitted by satellites. The signal from satellites located closer to the horizon line is delayed further. The delay error in the troposphere is approximately 0.5 m.

Another source of error is due to the satellite's orbit. Receiving devices calculate the coordinates of satellites in space using complex calculations, which consider the shape of the orbit and the speed of satellites, parameters that are not constant. The control segment monitors the location of satellites, calculates orbit eccentricity, and compiles this information into documents called ephemeris. The ephemeris is compiled for each satellite and transmitted along with the satellite signal. Receiving devices that have the capacity to process satellite ephemeris can compensate for some orbital errors. The error deviation of the satellite's orbit is approximately 2.5 m.

Satellites use atomic clocks that have high accuracy. Despite that, these clocks can suffer deviations of up to one millisecond, which is sufficient to reduce the accuracy of the positioning. These errors are minimized by the correction implemented by the system's monitoring stations. Receivers use quartz clocks that are less stable than the atomic clocks of the satellites. The receiver clock error is eliminated when it receives signal from one more satellite; however, other errors due to receiver noises will occur. The error due to the satellite clock is approximately 2 m, and the error due to receiver noises (clock and other causes) is approximately 0.3 m.

One of the factors that determine the positioning error of a GNSS is the way satellites accessible to the signal receiver are distributed in space. Receiving devices generally provide, along with coordinates, a parameter called position dilution of precision (PDOP). This parameter can have values between zero and infinity. PDOP is a qualitative information of the precision of the position measurement that depends on the position of satellites. The lower the value, the better the distribution of satellites for positioning. Ideally, PDOP should be less than 2, and it is generally not recommended to work when the PDOP is greater than 6. Thus, receivers that process signals from more than one type of GNSS (GPS, Galileo, and GLONASS) tend to have a lower PDOP, since they have a larger number of satellites to calculate the position and they can choose the ones that result in lower PDOP values.

The error of a GNSS can be expressed in several ways, and most of them are derived from the military science of ballistics. The most used are the circular error probable (CEP), the spherical error probable (SEP), the root mean square error (RMS), and the value equivalent to twice the root mean square error (2DRMS).

The circular error probable is defined as the radius of a circle on the earth's surface traced around the correct position of the point containing 50% of the determinations made by the receiving device. Thus, a CEP of 5 m means that 50% of the positions obtained are at most 5.00 m from the exact coordinate of the point to be determined. CEP is an error measure for horizontal positioning.

The spherical error probable is similar to the CEP with the difference that, instead of a circle, a sphere is used. Therefore, SEP is an error measure for positioning in three dimensions.

The root mean square error defines the positioning error of the receiving device with a 65% probability level. This means that, if the RMS positioning error is 5.00 m, 65% of the readings performed have an error of less than 5.00 m. On the other hand, a 2DRMS positioning error of 5.00 m means that 95% of the readings performed are up to 5.00 m from the exact coordinate of the point.

2.5   Systems to Improve the Precision/Accuracy of Satellite Positioning

Satellite positioning systems, without any kind of correction, can have positioning errors of up to 20 m. Due to the availability of various GNSS satellites, it is very common for a receiving device to receive the signal from 20 or more positioning satellites. Depending on the arrangement of the satellites, this error can be less than 3 m. With the modernization that is occurring in various GNSS systems, this error should be reduced to something around 1 m.

To reduce the GNSS positioning error, there are some alternatives. The first is to use a base station, located at a known coordinate point, to correct the error of the receiver used in the survey, which is usually called a rover. This system is called GNSS with differential correction.

Differential correction can be in real time or can be post-processed. To perform the correction in real time, it is necessary to have a communication system between the base and the rover. In the post-processed correction, both the rover and the base data are stored in files that will be processed later. Differential correction can be made based on the C/A code or based on the carrier phase.

When differential correction is made based on C/A code, a submetric accuracy is usually obtained, that is, an error less than 1 m. The system is called GNSS RTK when the correction is made using the carrier code and still occurs in real time; in the RTK system, the error is usually less than 10 cm. The rover in this case could be within a maximum of 6 miles (10 km) from the base station.

Another system that can be used is the GNSS Precise Point Positioning (PPP). In this system, the survey is carried out using only the rover. Positioning is obtained by correcting the errors associated to the satellite clock and satellite orbits and also to the effect of ionosphere on the GNSS signals. These errors are calculated by a base station network. This network is maintained by a service provider that centralizes the information in a control center, and this center sends the information to a geostationary satellite. The antenna of the rover, in addition to capturing the signal of the GNSS, also collects the information of this geostationary satellite.

As the increase in positioning accuracy is promoted by a satellite, this type of system is called Satellite-Based Augmentation System (SBAS). The SBAS can be regional or global. A regional SBAS has a geostationary satellite to increase the accuracy of positioning in a given region of the planet, which is a free service maintained by government agencies. Examples are the Wide Area Augmentation System (WAAS), which covers North America, and the European Geostationary Navigation Overlay Service (EGNOS), which covers the European continent. The global SBAS systems have one or more geostationary satellites, which increases positioning accuracy in all regions of the world, and this service is paid for and maintained by private companies. Examples are Starfire systems, maintained by John Deere; Starfix; OmniSTAR; and TerraStar. GNSS PPP provides real-time positioning. Generally, paid services provide more than one option in terms of correction, and the options that lead to a lower positioning error are obviously more expensive.

An important point to be defined by the user is what accuracy he needs to have in a positioning system. The operations that require the highest accuracy are those that rely on real-time signal for operations that involve the use of autopilot and control of machine traffic. For these conditions, the maximum error of the positioning system must usually be less than 20 cm. For operations involving the map-based variable-rate application of inputs, a 1 m error is often acceptable. For operations that involve the mapping of some soil, plant, or yield attribute, an error of up to 5 m is acceptable, since the data will need to go through an interpolation process that already involves a certain degree of error.

2.6   Final Considerations

In this chapter, an overview of the global navigation satellite system was presented. This is a very important topic in digital agriculture because almost all data that is generated needs to be georeferenced. The technology of GNSS is advancing in a fast way; every year, new systems are being created to improve the performance of the positioning systems. It is important that the professionals of digital agriculture stay updated on the innovations in this sector. We hope that with the information given in this chapter, the professionals that work in digital agriculture can be prepared to understand and use the technologies that are being developed.

Abbreviations/Definitions

- Global Navigation Satellite Systems (GNSS): General term used for any navigation satellite system.
- Global Positioning System (GPS): Global navigation satellite system implemented by the United States.
- Global Orbiting Navigation System (GLONASS): Global navigation satellite system implemented by Russia.
- Galileo: Global navigation system implemented by Europe.
- Navigation with Indian Constellation (NavIC): Regional navigation satellite system implemented by India.
- Quasi-Zenith Satellite System (QZSS): Regional navigation satellite system implemented by Japan.
- BeiDou: Navigation satellite system implemented by China.
- WGS84. World Geodetic Reference System used by GNSS to stablish the coordinates. Instead of using WGS84 is possible define any other local reference system, for example, SIRGAS2000. SIRGAS2000 is a geocentric reference system defined for Americas.
- Geoid: Surface that has equal gravitational potential. This surface is irregular.
- Ellipsoid: A model to represent the geoid defined by semi-major axis and flattening.
- Flattening: Dimensionless parameter that is obtained by dividing the difference between the semi-major axis and semi-minor axis by the semi-major axis.
- Universal Transverse Mercator (UTM). Map projection system. In the UTM system, the coordinates are given in meters units. The planet Earth is divided into 60 longitudinal projection zones with 6 degree longitude each that is numbered from 1 to 60. Each zone is converted to a flat surface.
- Geographic Coordinate System. The coordinates are given in latitude and longitude. The coordinates are expressed in degrees.
- L1: band transmitted by GPS at 1575.42 MHz.
- L2: band transmitted by GPS at 1227.60 MHz.
- Course Acquisition (C/A): Code obtained by signal modulation of L1 frequency signal, for civil use.
- Real-Time Kinematics (RTK): It is a differential correction system applied to GNSS. It is based on two GNSS receivers. One receiver is stationary (base station), and another receiver is used for collecting coordinates (rover). The base station sends the signal to rover to provide the corrections. The signal is sent by radio system, and the distance between base and rover should be less than 10 km. In RTK system, the error is typically less than 5 cm.
- Position Dilution of Precision (PDOP): Qualitative information of precision of the position measurement that depends on the position of GNSS satellites. The lower the value, the better the distribution of satellites for positioning. Ideally, PDOP should be less than 2, and it is generally not recommended to work when the PDOP is greater than 6.

- Circular Error Probable (CEP): Defined as the radius of a circle on the earth's surface traced around the correct position of the point containing 50% of the determinations made by the GNSS receiver.
- Spherical Error Probable (SEP): Defined as the radius of a sphere on the earth's surface traced around the correct position of the point containing 50% of the determinations made by the GNSS receiver.
- Root Mean Square Error (RMS): Defined as positioning error of the GNSS receiver with 65% probability level.
- 2DRMS: Equivalent value of twice of RMS Error.

Take Home Message/Key Points

- The satellite positioning systems has played an important role in the development of digital agriculture.
- The position in the geographic coordinate system is defined by two angles, latitude and longitude, and by the elevation relative to a reference surface.
- The geographic reference system is defined as the parameters related to the ellipsoid model. WGS84 is the optimized ellipsoid model for the world, and it is standard reference system for GNSS system.
- When a projection system is used, the position is projected onto a surface, and then that surface is transformed into a X,Y plane, while the point elevation represents the third coordinate.

References

Grewal MS, Andrews AP, Bartone CG (2020) Global navigation satellite systems, inertial navigation, and integration, 4th Edition. John Wiley & Sons, 521p
Van Sickle J (2020) Basic GIS coordinates, 3rd. edition. CRC Press, 171p

Chapter 3
Spatial and Temporal Variability Analysis

**Samuel de Assis Silva, Julião Soares de Souza Lima,
and Nerilson Terra Santos**

3.1 Introduction

In recent decades, agriculture has gone through a constant and irreversible upgrading process. Thus, great technological tooling has been employed, and the pursuit of greater efficiency in management techniques has been intensified, either by the use of modern equipment or by the search for agricultural practices that treat the field in a localized way.

Over time, the forms of management of agricultural areas have changed due to advances in agricultural sciences, based mainly on the concepts and techniques of precision agriculture (PA). By definition, PA can be understood as a branch of agricultural technology, based on the spatial and temporal variability of soils and crops, whose main objective is to manage crops in a precise and localized way.

The PA encompasses a set of tools and methods for the management of production fields, using different data collection platforms (sampling grids, sensors, satellites, unmanned aerial vehicles, and precise machines). Despite its comprehensive nature, PA is based on the management of spatial and temporal variability of phenomena that control agricultural production.

Unlike conventional agricultural models, PA not only makes use of values for a variable and/or attribute but also benefits from the variation of variables of agronomic interest over time and space. When considering the spatial and temporal variability of soils and crops, it is added, according to an agronomic understanding, a character of variation in micro and medium scales—that was not previously considered. Hence, the premise that managements based on average values are efficient

S. de Assis Silva (✉) · J. S. de Souza Lima
Universidade Federal do Espírito Santo, Alegre, Brazil

N. T. Santos
Universidade Federal de Viçosa, Viçosa, Brazil
e-mail: nsantos@ufv.br

D. Marçal de Queiroz et al. (eds.), *Digital Agriculture*,
https://doi.org/10.1007/978-3-031-14533-9_3

and capable of obtaining productive returns for the most diverse agricultural cultures is left aside.

Spatial and temporal variability is an inherent characteristic of all-natural phenomena, especially when considering dynamic and complex systems, such as soil-plant. Part of the variability observed in agricultural production fields can be considered intrinsic; however, a significant portion of this variability derives from the chaining of anthropic actions that modify the magnitude of variations and their behavior in space-time (Fig. 3.1). The emergence of the so-called digital agriculture came against PA's concept; this fact is due to the large volume of data—obtained by different types of sensors—that will contribute to the characterization of the spatial and temporal variability of the production fields.

Changes caused by poorly sized or inadequate agricultural practices can reduce the productive potential of the areas and culminate with results below the productive potential of cultivated plants. When analyzing the homogeneity in the fields of production, disregarding the existing variability, these effects tend to be more pronounced; fertilizations based on average values, for example, tend not to take into account the countless nutritional demands of the crops, which limits the physiological and productive capacities of plants.

Knowing the spatial and temporal variability of soil and plant attributes is indispensable to achieve excellence in agriculture. In this sense, it is necessary, as the first step in a PA program, to access this variability and determine its shape and magnitude; furthermore, it is necessary to understand how the combination of variability with other variables affects the development and yield of different crops. By establishing such issues, it is possible to subsidize the localized application of inputs, increasing the volume of information available for better agribusiness management.

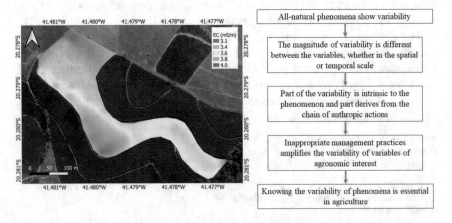

Fig. 3.1 Thematic map for the spatial variability of electrical conductivity (EC) and scheme to describe the impact of the variability in the agriculture. (Source: The authors)

3.2   Spatial and Temporal Variability Sources

As previously discussed, all-natural phenomena present variability, and each one of them presents a distribution characteristic and a different magnitude. Nevertheless, not all variations can be associated with a pattern in space and or time, being controlled, for example, by random factors. For the variation of a given phenomenon to be considered spatial, it must present a certain degree of spatial continuity or dependence (values of the nearest points are more similar than more distant points) of distribution between different positions (e.g., sampling locations) in the landscape. Phenomena that are distributed randomly are handled in agriculture, in most cases, based on average values, since the difficulty (or impossibility) of establishing distribution patterns prevents the adoption of localized management practices. For phenomena who are not random or do not behave as such, it is necessary to define the magnitude and extent of variability to generate maps that represent spatial variability accurately and recommend specific and localized management practices.

Regardless of whether a phenomenon presents random behavior or spatial and temporal dependence, the causes of these variations are often the same. Factors of landscape formation, along with geological processes, can be understood as the causes of natural variability, making up the structure of characterization of a variable and/or attribute, whose origin does not depend on another agent. In addition to natural sources, biotic and anthropic actions can induce variability and cause variations on smaller scales.

In the case of agriculture, the prominent variability is related to the soil-plant-atmosphere system; thus, the origins of these variations are linked to soil formation factors, plant genetic resources, and the effect of climatic variations. In this sense, and considering the complexity of the natural systems involved in an agricultural environment, it is expected that this variability complex will be enhanced by analyzing the interrelationships between the factors that control crop yield and by inserting anthropic action into the process. When considering soils, for example, the materials of origin, topography, and weathering are the main factors for the induction of natural variability, mainly spatial. The formation processes of agricultural soils are very variable, which resulted in a diversity of conditions and characteristics for this environment. Different soil classes present different physical, physical-chemical, and chemical characteristics (Fig. 3.2), which modifies the fertility of the areas and the ability/requirement of correction of essential factors from the productive point of view.

In addition to the differences between soil classes, due to the dynamism and interrelationships existing in natural environments, the variability also occurs in microscale, with large variations within the same class, mainly for chemical and biological attributes.

Among the soil attributes, the physical ones present the lowest spatial and temporal variability. In these attributes, continuity is more significant, and the effects of management practices are less impacting, with almost all of the observed variation coming from natural formation processes. On the other hand, physicochemical

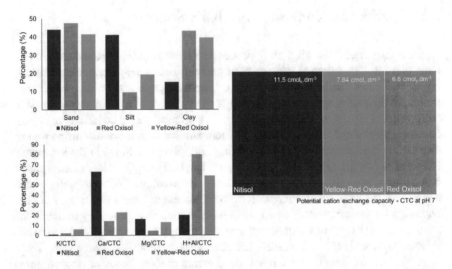

Fig. 3.2 Variation between physical (sand, silt, and clay), chemical (cationic ratios—K/CTC, Ca/CTC, Mg/CTC, and H + Al/CTC), and physical-chemical attributes (potential cation exchange capacity—CTC at pH 7) in different soil classes in cultivated areas. Among soil classes, the variation for physical, chemical, and physical-chemical attributes is high. Although it happens to a lesser extent, especially for physicists and physical chemists, the same class variation is also expressive. (Source: The authors)

Table 3.1 The scale of spatial and temporal variability of some soil attributes

Soil attribute	Variability	
	Spatial	Temporal
pH	Low	Low
Soil density	Low	Low
Total porosity	Low	Low
Silt	Low	Low
Clay	Mean	Low
Sand	Mean	Low
Cation exchange capacity	Mean	Mean
Base saturation	Mean	Mean
Exchangeable cations	High	High
Available micronutrients	High	High

Source: Adapted from Goderya (1998)

attributes show moderate variability, and the chemical ones are the elements that present the most significant variability, being the most influenced by management practices. Table 3.1 shows, in summary, the scale of spatial and temporal variability of some soil attributes.

For cultivated plants, variability needs to be considered when assessing the soil-plant-atmosphere system, given that climate changes and soil attributes, in addition to physiological issues, are decisive in the variability induction processes—mainly, in the temporal aspect. Depending on the botanical species, the cultivated variety, and the cultivation systems, the spatial and temporal variability observed for the plants can be greater or lesser, manifesting directly in the effective response.

Crop yield is, among plant attributes, the most studied one whose information is used to start a PA cycle. This fact denotes the importance of considering this factor in space-time, trying to extract information about a given area's variability. However, yield variability is canceled out when it is not related to other crucial agronomic information, not allowing a broad understanding of variability. Thus, it is vital to always associate the information on this variable's spatial behavior with the nutritional status (whether measured by traditional methods, like leaf analysis, or through remote sensing—proximal, aerial, and orbital) soil fertility and without disregarding environmental and climatic aspects.

When working to evaluate the variability of cultivated plants, it is essential to consider basic knowledge of morphology and physiology; this information is essential from data collection to decision-making in management practices. A clear example is the significant difference between working with plants of different sizes and different cycles (from annual to perennial). Shrubby plants, for example, have more significant intraplant variability than those of low vegetation; therefore, management should take these differences into account, increasing the complexity in interpreting the variability.

In general, in all situations, variability appears as obstacles to the development and yield of crops and opportunities for improving management methods and practices. In addition to natural factors, those with biotic and anthropic action alter the spatial and temporal distribution of individual variables, compromising crops' development and yield.

3.3   Factors That Affect Crop Yield

The spatial and temporal variability of yield is, among the plant attributes, the most studied in PA today. As variability is the central element for most crops (though in some cases, the quality of the product is more important than the volume produced), evaluating yield and its variation in space-time is decisive to initiate a broad diagnosis in the production fields. Despite the importance of yield variability, this variability must be related to the factors that control it, especially those whose effect is most decisive for each crop and cultivation system.

In an agricultural production system, factors hardly act in isolation; frequently, the elements act together, expressing variations that may go unnoticed if the spatial and temporal aspects are disregarded. Dampney and Moore (1999) classify the main factors that cause variability in agricultural crop yield according to their ability to change. The three major groups identified by the authors are the following: rigid (or fixed) factors, which are difficult to change and/or undergo alteration only by the action of phenomena unrelated to management practices; alterable (persistent) factors, which are amenable to change with management practices; and seasonal factors, whose change, in a short space of time, follows a particular behavioral pattern, depending on weather conditions like climate.

Rigid factors influence the development and yield of crops indirectly, given that their significant relationship is with other elements of the persistent type. In this group, most of the soil's physical characteristics can be listed; although important for all crops, these attributes are closely related to other elements that directly influence crop yield. An example of this interrelationship is given in the binomial phosphorus x soil texture. As an illustration, phosphorus has little mobility in the soil and close affinity with clay minerals. Generally speaking, clay minerals are responsible for making phosphorus available (Fig. 3.3).

For persistent factors, which comprise the soil's chemical attributes (macros and micronutrients), crop yield's effects tend to be directly evidenced; these effects are, however, rarely isolated. Generally speaking, the effects of spatial and temporal variability of soil attributes on crop yield happen through interaction between different attributes. Thus, it is always necessary to analyze the space-time behavior of more than one variable or attribute to search for the leading causes of yield variability (Fig. 3.4).

When evaluating the effect of spatial variability of attributes of soil and plants on the crop yield, it is necessary to consider, based on agronomic knowledge, the nutritional requirements of each cultivated variety, including spatial analysis as an additional source of information that allows a broad view of the production field. From this point, it is possible to establish distribution patterns and their effects on plant yield, identifying individual and interrelationship effects among attributes, guiding localized management practices aimed at managing variability.

Disregarding each crop's requirements and sufficiency indices is to ignore the real objective of spatial and temporal variability analysis and the objectives of PA. However, the method used is accurate and efficient, but the approach does not exclude basic agrarian science knowledge. Similarly, it is necessary to consider that observed spatial and/or temporal variability may not always result in an effect on yield variation. In several cases, despite the variability of the properties, the soil and

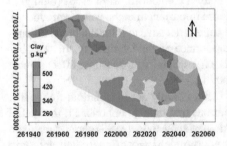
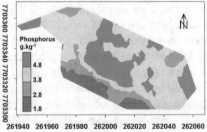

Fig. 3.3 Thematic maps of the spatial distribution of clay contents and phosphorus available in the agricultural production area. The spatial behavior of the two attributes is similar, indicating the direct relationship between them. Higher clay contents tend to contribute to the reduction of phosphorus availability due to adsorption processes, for example. (Source: Adapted from Silva et al. 2010)

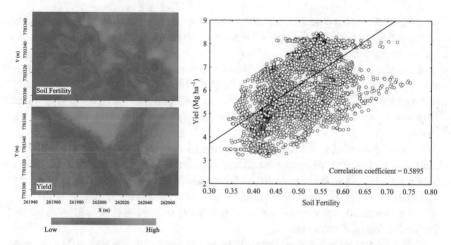

Fig. 3.4 The spatial relationship between soil fertility (defined by fuzzy modeling from the chemical and physical-chemical attributes that determine it) and conilon coffee yield. The spatial behavior of soil fertility is correlated with the yield spatial variability. Each input attribute for estimating the spatial variability of fertility has a different weight on the expression of yield, established according to agronomic criteria and nutritional requirements for each crop. (Source: Adapted from Silva et al. 2010)

plants may present values close to or higher than those suitable for plant development and yield, but they do not fit within the management range (Coelho 2003).

3.4 Temporal Variability of Crop Yield

As mentioned above, yield is an important response of most crops. Similarly, to other phenomena, crop yield presents variation in spatial and temporal planes controlled by various factors of the soil-plant-atmosphere system.

In PA, yield mapping is an essential process for decision-making in the management of agricultural areas. Despite their importance to the PA, yield maps are not always used correctly, mainly due to errors from the harvesters' monitors and the inability of farmers and technicians to analyze and interpret the presented variations.

With the availability of a large volume of crop yield data, often covering several years and/or cultivation cycles, mechanisms of analysis have been sought to understand the temporal variation and describe the phenomenon's behavior, evaluating its stability over different harvests. For a phenomenon to be considered temporally stable, it is necessary a constant combination in time between its spatial position and the statistical measures that characterize it (Wesenbeeck et al. 1988).

According to Blackmore et al. (2003), when crop yield is stable over time, it is possible to describe its distribution trend; consequently, it helps to predict spatial yield patterns for subsequent years, which would ultimately be a powerful tool for decision-making regarding agronomic management practices. On the other hand, when the behavior is unstable, it is not possible to establish a pattern of the temporal trend, and there is no possibility of predictions about its distribution and magnitude for future harvests.

To evaluate crop yield stability behavior, it is necessary to understand how this behavior varies along the spatial and temporal planes. In this sense, it is necessary to consider the fact that crop yield is not an independent variable; it is a direct response to a set of factors, which will be responsible for making it stable or not in time.

Generally speaking, the temporal variation of crop yield is associated with two main types of factors, called systemic and specific factors (Blackmore et al. 2003). Systemic factors can be characterized by variables or phenomena that act broadly and systematically, being distributed throughout the entire length of an area. In turn, specific factors are responsible for localized changes, implying variations in subareas within the production fields.

Systemic factors in which climate is the prominent example—since they act systematically—tend not to cause significant changes in the spatial distribution of crop yield (even if there are changes in absolute yield values). The maintenance of spatial patterns throughout different harvests is the primary condition for the temporal stability of crop yield.

According to Blackmore (2000), areas, where crop yield is temporally stable, are often more economically profitable than those where instability is predominant. Evaluating the temporal trend of crop yield in the United Kingdom, this author observed that the increase in stable areas' profitability was 15% than in the unstable areas.

The increase in the profitability of areas with temporal stability of crop yield may be associated with the possibility of optimizing agronomic practices and the adoption of localized management. Knowing the behavior of crop yield and predicting its distribution, it is possible, for example, to identify regions with lower yields caused by factors that cannot be circumvented through management, which can necessarily mean the reduction of the use of inputs (Blackmore 2000).

In the case of specific factors, the variation, since it is localized, tends to change the distribution of crop yield; hence, this variable's temporal instability will have a magnitude similar to the variability of such factors. Blackmore et al. (2003) state that in conditions where each year's variability is significant, there will be greater interannual instability, which prevents the following years' predictability. Under these conditions, the best strategy is to adopt management practices to manage variability within the same cultivation cycle.

It is worth reinforcing that, due to its dependence characteristic concerning other variables, the space-time variation of crop yield will always be related to their behavior. This fact, associated with agricultural production systems' complexity, is decisive for time instabilities to be expected for different cultivation types. Despite this fact, the information on crop yield and its space-time variation cannot be neglected in the procedures for identifying homogeneous management zones.

3.5 Determination of Spatial and Temporal Variability of Soils and Crops

As previously discussed, for the variation of a given phenomenon to be considered spatial and/or temporal, it must present continuity of distribution between different positions in the landscape or in time. When it happens, it is possible to affirm that this phenomenon is dependent, and we can, from specific methodologies of analysis, describe it in terms of magnitude and scope in space-time.

The description of the variation of different variables is usually performed using traditional methods of statistical analysis, such as variance, standard deviation, and coefficient of variation. For phenomena with continuity throughout space and time, however, these statistics are not the most recommended, since they consider that variations from one place to another are random, ignoring the relationship between them as a function of the distance (in time or space) that separates them.

For continuous phenomena in space and time, geostatistics is one of the recommended methods for data analysis. Through autocorrelation and semivariance functions, it is possible to determine the current spatial and temporal dependence for the same variable—in terms of the positional variation where it is located (Oliver 2010).

Geostatistics is based on the theory of regionalized variables, which recommends that a variable be distributed in space or time; thus, it can be seen as a realization of spatially correlated random variables. In a context of regionalization, it is expected that the similarities to values of the same variable—measured in different positions in the spatial or temporal plane—decrease as the distance between observations increases.

Starting from the premise that "two random variables are independent when the covariance between them is equal to zero," geostatistics assumes that (a) the covariance between different measures of the same variable decreases as the distance between two reference samples (tending to zero) increases—in this context, higher covariance values are indicative of greater autocorrelations and (b) the semivariance between different measures of the same variable increases as the distance between two reference samples is increased (tending to equality with the data variance)—in this context, lower values of semivariance are indicative of greater autocorrelations.

To define the structure of spatial and temporal dependence of different variables, based on the hypothesis of regionalization, geostatistics uses two main methods: the correlogram and the variogram (also known as semivariogram). In both methods, the phenomena or variables must be stationary; thus, they need to present some homogeneous behavior along the spatial and or temporal plane (Oliver 2010). Despite the standard requirement of stationarity, the variogram is less restrictive (intrinsic hypothesis) and, therefore, more used to describe phenomena' spatial and temporal variability.

Correlograms are recommended for second-order stationarity, that is, finite variance and constancy in average and covariance values. The correct description of the spatial and temporal variability of a variable permeates the choice of correct data analysis methods and, consequently, the analyst's knowledge and skill. However, this process requires that all stages of the procedure be performed correctly, starting from an efficient and representative sampling until the interpretation of results and the decision-making for the variability management (Fig. 3.5).

The sampling step for data collection is the first within a flowchart for spatial analysis. Data to be submitted to geostatistical procedures must necessarily present coordinates (geographic or not) in space and/or time. In this sense, this stage is of paramount importance and can be considered one of the most critical points in this process, given all the implications involved, such as high density, costs, and sampling methods, among others.

From a correct sampling, it will be possible to establish the scale of evaluation, analysis, and interpretation of spatial analysis results. It is essential to note the sample density and spacing among samples.

Webster and Oliver (1992) claim that the Matheron's classic variogram requires a minimum of 100 sampling points to reflect reliable and accurate estimates. Despite this fact, not always 100 sampling points will be sufficient or required to describe the spatial and/or temporal variability of a phenomenon. For this to happen, it is necessary to consider factors such as the mapping's objectives, the scale or dimension of the work area, the nature of the work, and, mainly, the type of variable to be analyzed.

It is worth mentioning that the procedures for generating the database for spatial and temporal variability analysis are those that most encumber the systems in PA. Samplings and the laboratory analyses do not always present a cost that makes them possible, which is the primary source of the obstacles to the correct adoption of reliable geostatistical models.

Once the database is built, the next step is exploratory statistical analysis. At this stage, the procedures consist of determining the measures of position, dispersion, and its shape, in addition to the relationship and correlations among variables. Although basic, these analyses are necessary because they offer an overview of the variables under study, allowing the user to find anomalies in the data set, understand the distribution and average behavior, determine interdependence or relationships between different variables, and identify distribution trends.

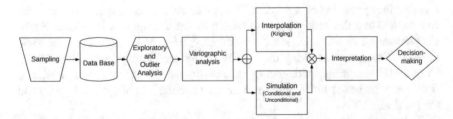

Fig. 3.5 Generic flowchart of the spatial analysis process. The stages presented are generic; however, they present a rational sequence of procedures for collecting, analyzing, and interpreting specialized data aiming at the management of variability. (Source: The authors)

Ultimately, these statistics help to understand the spatial and temporal distribution behavior of variables, because when behavior patterns are described (even without considering the sample position), they guide the interpretation of the results and the need, or not, for further analysis.

Although necessary for the data analysis process, classical statistical methods, as mentioned above, are ineffective in describing the spatial and temporal variability of different variables, since they do not consider the position in space and/or time of the samples of the same variable. In conditions where the variation is evident with some degree of organization, according to the position in the spatial and/or temporal plane, geostatistics should be used to describe this continuity and the degree of dependence among the samples.

In the geostatistical analysis, the main objective is to describe the spatial or space-time continuity of a phenomenon. Using geostatistical methodologies, we advance in the studies and description of a variable, leaving the limitation of conventional methods that simply present a discrete and limited summary of the phenomena, aiming to understand the influence of space and time on its variability.

The variogram is the main geostatistical tool for describing the spatial or space-time continuity of a variable. This tool allows us to graphically demonstrate the structure of the variability, characterizing it from different parameters and quantifying the degree of dissimilarity between observation pairs separated by the distance vector. In this sense, it is not incorrect to state that the variogram is the best tool to represent a regionalized phenomenon; however, its efficiency is influenced by sampling geometry.

Equation 3.1 presents the model of Matheron's classic variogram estimator. The equation presents the function of semivariance [$\gamma(h)$], which measures the degree of dissimilarity among samples of the same variable. The semivariance grows as the distance between the pairs of sample points increases, tending to the data variance value, from where the behavior of the variable loses the character of regionalization and becomes random.

$$\gamma(h) = \frac{1}{2N(h)} \sum_{i=1}^{N(h)} \left[z(u_i) - z(u_i + h) \right]^2 \tag{3.1}$$

In general terms, the variogram is a function dependent on the distance (h) among samples, that is, the distance vector (modulus and direction) among specialized sample pairs [$N(h)$] specialized [$z(u_i) - z(u_{i+h})$]. As already reported, the larger the module of vector h, the greater the dissimilarity between the pairs of points analyzed, increasing the value of the semivariance (Fig. 3.5). Each point in Fig. 3.5 represents a semivariance, calculated for a distance h, according to Eq. 3.1. The plotted points in Fig. 3.6, as a function of the distance h is the experimental variogram. The curve that passes through the points is known as the theoretical variogram.

The semivariance does not start at the origin of the coordinate axes; it presents a discontinuity (nugget effect—C_0), corresponding to microscale variability and, in some cases, to sampling errors. In a variographic analysis, the semivariance

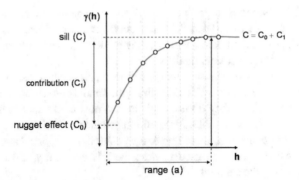

Fig. 3.6 Classical theoretical variogram, with an indication of the parameters estimated from the geostatistical analysis. The graph contrasts the semivariance [$\gamma(h)$] values for each h distance. The semivariance grows until it reaches the sill (C) where it stabilizes. The distance at which the semivariance reaches the value of the level is the range (a) of the semivariance, in which the variable's behavior is regionalized. (Source: Authors)

calculations are performed in different directions—usually quadrants—(0°, 45°, 90°, and, 135°) of vector h, seeking to detect whether the phenomenon is isotropic or anisotropic. Isotropy consists of similarity of spatial behavior in all directions analyzed, in which a single mean variogram is able to describe the distribution of the variable fully. In the case of anisotropic behaviors, it is necessary to detect the directions of greater and lesser spatial continuity of the investigated phenomenon.

In addition to the questions related to anisotropy, some natural phenomena show a distribution trend (usually evidenced in variograms with increasing semivariance and undefined sill) or interlaced structures, in which more than one sill and/or more than one range are observed in the analysis. For the latter case, it is possible to affirm that there are different ranges of variability for the same variable within the same sample universe.

Regardless of the obstacles discussed earlier for variographic analysis, variograms can be described as a set of pairs of discretized points. For these discrete positions, continuous functions capable of describing small variations in the semivariance × distance (h) ratio are adjusted.

These continuous functions are called models and represent an average attenuated curve representing the variable's spatial continuity. The adjustment of the theoretical models to the experimental variograms is a crucial step in the geostatistical analysis; from these models, the parameters of the variograms are determined, being the same used in the procedures of interpretation geostatistical estimation.

Several models or functions can be used to model (describe) the experimental variogram; however, the choice should be based on those that are able to provide stable solutions and reduced errors in the subsequent processes of geostatistical estimators. Within the available models, the most common are as follows: (a) spherical—more usual in variables of the soil-plant system, effectively achieving stabilization when the semivariance tends to match the data variance; (b) exponential—quite frequent in agronomic variables, it is also a function of the threshold and reach;

however, this first parameter is reached asymptotically; and (c) Gaussian—unlike the two previous models, it presents a slow growth of semivariance over shorter distances, being frequent for variables that present greater regularity and continuity of distribution; similarly to that observed for the exponential model, it tends, in an asymptotic way, for the variance of the data.

The choice of model should be judicious to ensure greater accuracy in the estimates of variogram parameters. Generally, the procedures for selecting the models that best fit the experimental variograms are based on the least squares theory, in which the models that present the lowest values for the sum of squared residues (SSR) are selected and also higher value for the coefficient of determination (R^2).

The least squares method is exclusively related to the adjustment of the theoretical model's continuous curve to the experimental model; this method is not efficient in evaluating the accuracy of estimating variogram parameters on the quality of interpolations. For these cases, cross-validation is a vital methodology to be considered when choosing models. Cross-validation expresses the relationship between values observed and estimated by geostatistical linear interpolators, presenting different error statistics and assisting the choice of theoretical models that minimize deviations in interpolation (Fig. 3.7).

After choosing the model that best describes the experimental variogram's behavior, the later procedures aim to represent, continuously in space and or in time, the distribution behavior of the phenomenon or variable. The geostatistical linear interpolation and stochastic simulation methods are the most common to describe, from thematic maps, this continuous representation in the spatial and temporal planes.

In geostatistical interpolation, the objective is to estimate values for unsampled sites, with the minimum estimation error variance. The method used in this process is kriging (and its variations). This method is currently considered one of the most

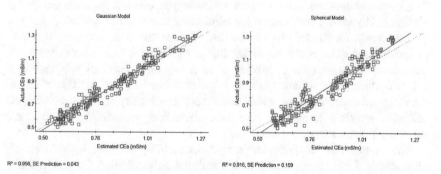

$R^2 = 0.956$, SE Prediction = 0.043 $R^2 = 0.916$, SE Prediction = 0.159

Fig. 3.7 Cross-validation graphs for different theoretical models adjusted to the experimental variogram of the soil's apparent electrical conductivity (mS.m^{-1}). The correct choice of the model reduces the quadratic error of interpolation by ordinary kriging. The Gaussian model produced an average quadratic error of 0.043 mS.m^{-1} in the estimation of the soil's apparent electrical conductivity, while for the spherical model, the mean error was 0.159 mS.m^{-1}. (Source: Authors)

accurate for estimating unmeasured values, mainly because it considers the structure of spatial variability determined from the theoretical variograms.

Since it is a geostatistical approach, kriging, unlike other methods, only applies to data that exhibit spatial dependence behavior. Likewise, a few methods present interpolation uncertainty because it estimates values for the sampled positions (actual data), comparing the interpolated results with those observed. It allows the user to decide on the methodology's accuracy, using fundamental knowledge about the behavior of the variable under study.

Like all interpolation methods, kriging presents a simplification of reality, since it reproduces the most likely form of distribution of a phenomenon/variable (Soares 2006). When the goal is to maintain "fidelity" concerning spatial variability and measured actual values, simulation is an essential tool because it provides a set of equiprobable scenarios for the variable, maintaining its spatial distribution behavior. Since it is not the objective of this chapter, for further studies on geostatistical simulation, we recommend the reading of Lantuéjoul (2002).

The method's choice for describing spatial variability in thematic scenarios should include the study's objectives and needs and/or application. Frequently, estimation methods will be preferred (interpolation), since agricultural sciences are correctly described in the products generated by these approaches—with no underestimation or overestimation that compromises the interpretation of results and decision-making. However, as Soares (2006) warned, when the smallest dispersion of the values estimated by the interpolation methods generates distortions in the thematic scenarios (mainly in the extreme values), simulation should be considered in substitution to the estimation in order to minimize interpretation bias.

3.6   Management of Variability

The studies of the spatial and temporal variability of soils and crops are usually not carried out to provide information for eliminating the variation existing in the production fields. As already discussed, many of these variations originate in natural processes, and when evaluating an infinity of variables/attributes, it would be presumptuous to imagine the possibility of reversing such processes with anthropic practices during one or more cultivation cycles. Therefore, the objective of understanding and determining the spatial and temporal variability of the attributes of the soil-plant system is to manage these elements, making agronomic practices more efficient, individualized, and localized.

The management of spatial and temporal variability of the soil-plant system's attributes is, from an agronomic or horticultural point of view, the practical and direct application of the entire process of analyzing the magnitude and scope of the concept of variability. Transforming the information obtained from geostatistical analysis into useful insights for the management of production fields is the initial part of a process that involves other PA tools; these tools are responsible for site-specific treatments of crops.

Despite the importance of studying spatial and temporal variability for a broad understanding of production fields, it is necessary to establish cause and effect relationships among different sets of variables. However, it is worth noting that it is not always possible to establish such relationships; nonetheless, it is vital to seek information to understand the variability of a given effect (e.g., crop yield), which allows guiding the adoption of more efficient management strategies.

Thanks to the advancement of PA and information technologies (through collecting, processing, storing, and analyzing a large amount of data from different platforms and methods), the use of integrated systems for decision-making in agricultural production fields has become increasingly common. Information on the variability of soils, crops, and the behavior of pathogens and agricultural pests are—when analyzed with other variables of agronomic interest—powerful tools for decision-making.

Regarding the information related to a set of different variables, the most usual way to manage variability, mainly the spatial one, is through management zone delineation. Using specific methodologies for grouping data, it is possible to subdivide the fields into "uniform" classes for different attributes and adopt specific management approaches within each class. Considering the existing heterogeneity in the fields of production and the crop management from homogeneous subunits, management is carried out in a targeted manner, so that excesses or deficits are reduced. In this division into management zones, the following aspects can be considered: (a) spatio-temporal variability of soils, (b) climate, (c) plants, (d) the performance of agricultural machinery and inputs, (e) issues related to local geography, and (f) environmental safety.

Finally, spatial and temporal variability is a characteristic of soils and crops and needs to be considered in agronomic management practices. However, in studies and practical applications, it is necessary to consider the real effects of variables/attributes on the crop yield to make the processes in precision agriculture more efficient.

For an accurate characterization of the magnitude and scope of variation in space and time, all stages of the analysis process must be carried out carefully, ensuring that the processing results are explicable and do not compromise the subsequent analyzes for making decision-making. The analysis of variability involves the improvement of knowledge in spatial analysis techniques—particularly geostatistics—and must be performed respecting the basic assumptions of the methods, from data collection planning to the choice of the variograms models.

The joint use of a significant amount of information collected periodically can increase the efficiency of management practices. This method makes it possible to understand the conditions of crops and cultivation systems during several production cycles, offering a global view of the area and subsidizing decision-making to manage variability.

Abbreviations/Definitions

- Geostatistics: It is a mathematical language for the study, characterization, and modeling of regionalized variables, whose distribution is spatially continuous. From autocorrelation and semivariance functions, geostatistics allows determining the existing spatial dependence for the same variable as a function of the variation in different between positions sampling.
- Variogram: Geostatistical function that describes the structure of the spatial variability of a variable, characterizing it from different parameters and quantifying the degree of dissimilarity between observation pairs separated by the distance vector (h).
- Regionalized variables: Variables distributed in space or time, defined as a realization of a set of spatially correlated random variable.
- Anisotropy: Characteristic of a regionalized variable or phenomenon, in which the spatial distribution is different in different directions (quadrants) in a spatial plane.
- Kriging: Linear geostatistical method of interpolation. Estimates values for unsampled locations from the information from adjacent samples and based on the spatial dependence structure determined by the variogram. It is considered an unbiased interpolation method with minimal variance.
- Stationarity: In geostatistics, it is said of the property of phenomena or variables in relation to their more or less homogeneous behavior along the spatial and/or temporal plane.
- Nugget effect (C0): Corresponding to microscale variability and, in some cases, to sampling errors. It is the discontinuity at the origin of the coordinate axes, with the semivariance does not start at the origin.
- Sill (C): It is the position in axis of the semivariance where the curve of the theorical variogram becomes constant.
- Range (a): It is the position in axis of the distance at which the semivariance reaches the value of the level, in which the variable's behavior is regionalized.
- Cross-validation: It expresses the relationship between values observed and estimated by geostatistical linear interpolators, presenting different error statistics.

Take Home Message/Key Points

- All-natural phenomena show variability and the magnitude of variability is different between the variables, whether in the spatial or temporal scale.
- Part of the variability is intrinsic to the phenomenon but the inappropriate management practices amplifies the variability of variables.
- The geostatistics analysis are used for the accurate characterization of the magnitude and scope of variation—in space and time—of variables.
- Knowing the spatial and temporal variability of soil and plant attributes increases the efficiency of agronomic management practices.

References

Blackmore S (2000) The interpretation of trends from multiple yield maps. Comput Electron Agric 26:37–51

Blackmore S, Godwin RJ, Fountas S (2003) The analysis of spatial and temporal trends in yield map data over six years. Biosyst Eng 84(4):455–466

Coelho AM (2003) Agricultura de Precisão: manejo da variabilidade espacial e temporal dos solos e das culturas. In: Curi N, Marques JJ, Guilherme LRG, Lima JM, Lopes AS, Alvarez V (eds) V.H. Tópicos em Ciência do Solo, vol 3. Sociedade Brasileira de Ciência do Solo, Viçosa, pp 249–290

Dampney PMR, Moore M (1999) Precision agriculture in England: current practice and research-based advice to farmers. In: Robert PC, Rust RH, Larson WE (eds) Precision agriculture. ASA, CSSA, SSSA, Madison, pp 661–673

Goderya FS (1998) Field scale variations in soil properties for spatially variable control: a review. J Soil Contam 7:243–264

Lantuéjoul C (2002) Geostatistical simulation—models and algorithms, 1st edn. Springer, Berlin/Heidelberg, 256p

Oliver MA (ed) (2010) Geostatistical applications for precision agriculture. Springer, pp 1–34

Silva SA, Lima JSS, Souza GS, Oliveira RB, Silva AF (2010) Variabilidade espacial do fósforo e das frações granulométricas de um Latossolo Vermelho Amarelo. Rev Ciênc Agron 41:1–8

Soares A (2006) Geoestatística para as ciências da terra e do ambiente, 2nd edn. IST Press, Lisboa, 214p

Van Wesenbeeck IJ, Kachanoski RG, Rolston DE (1988) Temporal persistence of spatial patterns of soil water content in the tilled layer under a corn crop. Soil Sci Soc Am J 52:934–941

Webster R, Oliver MA (1992) Sample adequately to estimate variograms of soil properties. J Soil Sci 43:117–192

Chapter 4
Images and Remote Sensing Applied to Agricultural Management

Flora Maria de Melo Villar, Jorge Tadeu Fim Rosas, and Francisco de Assis de Carvalho Pinto

4.1 Introduction

The growth in the world population and the change in people diet demanding for more protein-rich animal foods is causing an increase in the demand for grain and oilseeds. China, the United States, India, and Brazil are the main world producers and are playing an important role in supplying food for different parts of the world. Considering that all over the world, new areas for agricultural production are becoming scarce, to meet challenge of increasing the production at a level that attends the world's demand is necessary to find ways of improving the efficiency of how food is produced. For sure, digital agriculture tools are having an important role to meet the food demand of the world population.

In recent years, aiming at the sustainable development of agriculture, many tools have been used to monitor agricultural fields. The monitoring of agricultural crops helps the decision-making process, ensuring better management of resources. There are several methods of this monitoring; however, remote sensing has stood out due to the ease of acquiring information in different extensions of areas and in a short period of time, facilitating the characterization of spatial and temporal variability of agricultural production fields. In addition to assisting in the management of the use of resources by producers, the adoption of remote sensing allows the techniques of this tool to be applied more efficiently for estimating production, estimating cultivated areas, detecting the effect of droughts, and identifying pests and diseases in crops.

F. M. de Melo Villar (✉) · F. de Assis de Carvalho Pinto
Federal University of Viçosa, Viçosa, Brazil
e-mail: flora.villar@ufv.br; facpinto@ufv.br

J. T. F. Rosas
University of São Paulo – 'Luiz de Queiroz' College of Agriculture, Piracicaba, Brazil
e-mail: jorge.fimrosas@usp.br

© The Author(s), under exclusive license to Springer Nature
Switzerland AG 2022
D. Marçal de Queiroz et al. (eds.), *Digital Agriculture*,
https://doi.org/10.1007/978-3-031-14533-9_4

4.2   Remote Sensing

All bodies with temperature above absolute zero degree (0 K = −273 °C) emit electromagnetic energy. Remote sensing can be understood as the technique that uses sensors, without direct contact with the object of interest, to obtain information by capturing its reflected and/or emitted electromagnetic radiation (EMR). For this, a remote sensing system consists basically of a source of electromagnetic radiation (e.g., the sun), a terrestrial target (vegetation, water bodies, soil, urban areas, etc.), a platform that carries sensors and a site of data processing, storage, and distribution. The sensors, responsible for capturing electromagnetic energy from terrestrial targets, can be imaging, providing an image of the observed area as a product, or non-imaging, providing data in graphical or numerical form as a product. Photographic cameras or scanners and radiometers or spectroradiometers are examples of imaging and non-imaging sensors, respectively (Formaggio and Sanches 2017).

EMR comprises ranges of the electromagnetic spectrum, which in turn can be defined as the set of all existing electromagnetic waves. The bands of the electromagnetic spectrum do not have well-defined exact limits, but we will address here limits that are pertinent to several fields of the literature. In remote sensing, there are equipment and devices available to explore from the spectral band of ultraviolet (wavelength from 0.3 to 0.4 µm) to microwaves and radio waves (above 27,000 µm, which is equal to 2.70 cm) (Fig. 4.1). The visible spectrum (0.4–0.7 µm) is the range that the human visual system can perceive, in which each wavelength provides a different sensation of color, from violet to red. In a very simplified way, it is common to divide the visible spectrum into three bands of wavelengths: blue (0.4–0.5 µm), green (0.5–0.6 µm), and red (0.6–0.7 µm), called RGB (red, green, blue). The infrared spectrum (0.7–15.0 µm) is divided into near infrared (0.7–1.3 µm), medium infrared (1.3–5.6 µm), and far infrared (5.6–15.0 µm). It is also common to divide the infrared spectrum into reflected infrared (0.7–3 µm) and thermal or emitted infrared (3–15.0 µm).

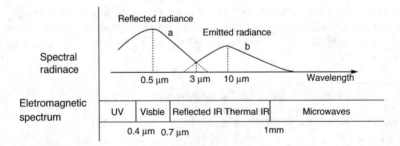

Fig. 4.1 Radiance reflected and emitted considering the spectral regions of a given object. The sensors used for data acquisition for the visible range are cameras that comprise the visible range; photodetectors are used for thermal infrared, and microwave sensors are used for the microwave region. (Source: Adapted from Formaggio and Sanches 2017)

When it comes into contact with a surface the radiation can be absorbed, transmitted, and reflected. The spectral variable reflectance, obtained through the ratio between the flux of radiation incident on a surface and the flux of radiation that is reflected by it, is a property of the object widely used in remote sensing. On most natural targets on the Earth's surface, the reflectance for each wavelength is not the same (Fig. 4.2). This characteristic specific to each target on the Earth's surface is called spectral signature and corresponds to the reflectance curve along the electromagnetic spectrum.

In agricultural fields, knowing the spectral characteristics of the canopies can contribute to the detection of normal or abnormal patterns that tend to occur in cropping areas. The spectral signature of the vegetation follows a certain pattern (Fig. 4.2). The vegetation tends to reflect more radiation in the near infrared region (0.7–1.3 μm) due to the physical structure of the leaves. Vegetation absorbs virtually no energy in the near infrared region, and the portion that is not reflected is transmitted. On the other hand, vegetation tends to absorb most of the radiation in the visible spectrum band (0.4–0.7 μm), mainly red and blue wavelengths, because these radiations are related to leaf pigments responsible for the photosynthetic process. In the medium infrared region (1.3–5.6 μm), the spectral behavior is influenced by the

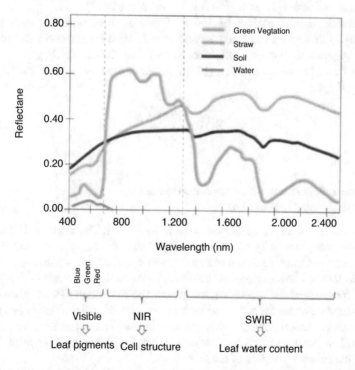

Fig. 4.2 Example of spectral signatures of targets within the range from 400 nm to 2400 nm. Visible, wavelength bands from 400 to 700 nm; NIR, near infrared wavelength bands (700–1300 nm); and SWIR, shortwave infrared wavelength bands (1300–2500 nm). (Source: Adapted from Formaggio and Sanches 2017)

radiation absorption peaks of the leaf water molecule (1.4 μm, 1.9 μm, 2.5 μm, and 2.7 μm).

Chlorophylls are responsible for the green color of vegetation because they have radiation absorption peaks in the blue and red regions of the spectrum. The other photosynthetic pigments have these absorption peaks only in the region of the blue spectrum. This fact alone makes the spectral band of red the most used in remote sensing compared to the others in the visible spectrum, for example, in the equations of vegetation indices.

Basically, there are two ways of analyzing the remote sensing images. One is by using the reflectances of the different bands in the image, and the other is by creating what is called vegetation index. The problem of using the reflectances directly is that for each pixel you have more than one data, making difficult to understand the spatial and temporal variability in the area that is being analyzed. The vegetation indices are created to maximize the sensitivity that the obtained remote sensing data has related to one or more vegetation characteristics. For instance, you can design a vegetation index that is sensitive to the amount of biomass in the vegetation or to presence of a disease in the area. The other advantage of using vegetation index is to minimize the effect of confounding factors such as background reflectance and directional and atmospheric effects (Liang and Wang 2019).

Vegetation indices are combinations of two or more reflectances in specific wavelengths or spectral bands to generate an index that is related to a characteristic of the vegetation. For example, the Normalized Difference Vegetation Index (NDVI), described by Rouse et al. (1973), is one of the most popular indices and is given by Eq. 4.1.

$$NDVI = \frac{NIR\,?\,R}{NIR + R} \tag{4.1}$$

where

NIR is the reflectance in the near infrared band.
R is the reflectance in the red band.

NDVI assumes values within the range from −1 to 1. The closer to 1, the higher the photosynthetic activity of the vegetation at the site. Negative or near 0 values indicate areas of water bodies, buildings, exposed soil, or places where there is little vegetation, that is, low chlorophyll activity. The theoretical principle of this index is that the more active the vegetation, the greater the absorption of sunlight in the red region and the greater the reflection in the near infrared region. This is due to photosynthesis and chlorophyll activity, resulting in low values of reflectance in the red channel. The greatest reflection in the near infrared region is due to physical factors, such as the cellular structure of the leaves.

Once the spectral behavior throughout the phenological cycle of crops is known, any observed alteration may indicate an anomaly or a natural change that is part of the plant cycle. For example, localized herbicide spraying sensors, such as Trimble's

WeedSeeker®, locate weeds based on the difference of NDVI between soil and vegetation.

The acquisition of EMR reflected by the target in remote sensing is done through optical sensors, which in turn are divided into passive sensors and active sensors. Passive sensors are defined as those that use sunlight as a source of radiation at the time of data acquisition, whereas active sensors use artificial sources of radiation. Photographic cameras are examples of passive sensors, while NDVI meters such as Trimble's GreenSeeker® and Holland Scientific's Crop Circle ACS-211® can be listed as active sensors.

Optical sensors are coupled to platforms, as mentioned earlier. Platforms for data collection can be generally classified into orbital (satellite), aerial (high, medium, and low altitude), and terrestrial (field or laboratory) levels (Fig. 4.3).

4.3 Digital Images

In the context of remote sensors, digital cameras have been widely used in recent decades, mainly on aerial platforms. Digital cameras have a photosensitive sensor, usually a charge-coupled device (CCD) or a complementary metal-oxide-semiconductor (CMOS). These sensors are subdivided into small cells arranged in matrix format. As it comes into contact with the EMR, each sensor cell generates an amount of electrons as a function of the amount of energy that has reached it. Thus, electronic systems with integrated camera software are able to read and transform the amount of electrons into a digital number that is directly correlated with the amount of energy that reaches the photosensitive cell.

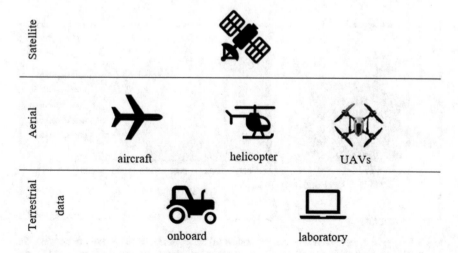

Fig. 4.3 Data collection levels: orbital, aerial, and terrestrial

The information recorded by digital cameras is stored in numerical matrices that are called digital images. In practical terms, each element of the matrix that forms an image is called a pixel, which is presented on a scale of integers greater than or equal to zero, whose number of levels is a power of two and depends on the radiometric resolution of the sensor. So, for instance, an image with 8-bit radiometric resolution has 2^8 (equal to 256) levels of numerical values of pixel, ranging from 0 to 255, while a 12-bit image has 2^{12} (4096) levels, ranging from 0 to 4095.

Each spectral band can be captured by placing an optical filter in front of the sensor, allowing only the wavelengths of interest to pass through and reach the sensor. In this case, either the energy enters the camera only through a lens and, internally, is diverted to each sensor, or the energy enters independently for each sensor. For instance, a digital camera with five spectral bands with independent imagers for each sensor is shown in Fig. 4.4.

However, due to more affordable costs, color digital cameras (three bands—R, G, and B) have been often investigated for information acquisition in agriculture. Unlike the camera in which each band has a filter-sensor set, this camera has only one sensor, with the optical filters of the three bands glued to each cell individually, in an array known as the Bayer Filter (Fig. 4.5). The electronic camera system is responsible for the individualization and formation of each R, G, and B band. In view of the importance of the near-infrared spectral range in the sensing of vegetation, these cameras that employ the Bayer filter have been modified for acquisition of this range, for example, by changing the filter that blocks the visible range to one that allows the passage above the wavelength band of red. However, the three bands that are generated after the modification are sensitized by the energy in the red and infrared bands, making it difficult to isolate each band. These cameras have the disadvantage of not enabling the configuration of each channel independently.

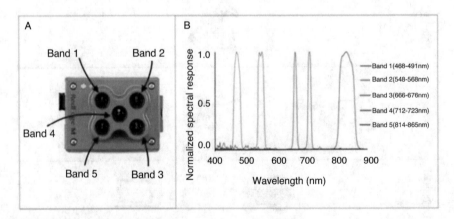

Fig. 4.4 Digital camera with five spectral bands: (**a**) arrangement of the sensors responsible for the acquisition of each band individually; (**b**) spectral sensitivity of each band recorded by the multispectral camera. (Source: Adapted from Wang and Thomasson 2019)

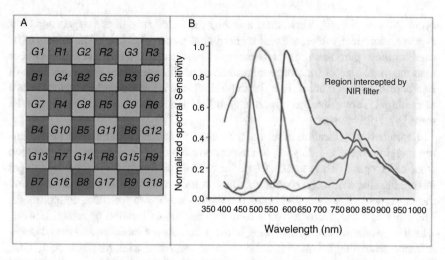

Fig. 4.5 Digital camera with one RGB sensor: (**a**) Matrix pattern of photosensitive cells of the filter Bayer and (**b**) spectral sensitivity of each cell. (Source: Adapted from Lebourgeois et al. 2008)

The numerical value of each pixel of the image is a function of the sensitivity of the sensor (Figs. 4.4b and 4.5b), the illumination that reaches the target and the spectral signature of the target (Fig. 4.2). However, in order for decisions based on the information generated by remote sensing to have higher probability of success, it is necessary to use the reflectance of the target, not the numerical value of the pixel. The transformation of the numerical value of the pixel into reflectance is performed by radiometric calibration. If, on the one hand, in the orbital remote sensing images are often already made available after radiometric calibration, in aerial sensing, this calibration has not been a reliable operation, especially with the widespread use of the modified Bayer filter cameras, given the overlap between spectral bands (Fig. 4.5b).

The most commonly used calibration method for radiometric calibration of aerial images is the empirical line method. This method assumes that the reflectance of surfaces has a linear relationship with the numerical value of the image. Thus, reflectance can be estimated by converting the numerical value of the pixel through a simple linear function (Eq. 4.2).

$$R_i = aV_i + b \qquad (4.2)$$

where

R_i is the reflectance of pixel in band i, %.
V_i is the numerical value of pixel in band i, dimensionless.
a and b are the parameters fitted by linear regression.

To determine the parameters of the calibration equation, this method makes use of targets with known reflectance present in the study area at the time of image acquisition (Fig. 4.6b). Thus, after knowing the reflectance of the targets and their average numerical values, a linear regression is made (Fig. 4.6c). The targets to be used for these purposes must be homogeneous, must have a size a few times larger than the pixel size in the image, and must have characteristics similar to those of Lambertian bodies. That is, in order to be similar to a Lambertian body, the surface of the target is considered to be perfectly diffuse and radiance is considered to be the same in all directions.

Radiometric calibration of orbital sensors is performed differently because, during imaging, there is influence of atmospheric factors, temperature, and radiation from outer space (Zhou et al. 2015). On orbital platforms, instead of the imager having the shape of a cell matrix, the most common is that it is a line of cells, and the movement of the satellite scans the Earth's surface to form the digital image. There are different methods to perform radiometric calibration of orbital sensors, which have been changing according to the generation of satellites and the advancement of technology. On-orbit calibration of satellites with MODIS, ALI, OLI, ETM+ sensors, etc., requires the calibration of solar radiation or calibration of lamps, which are the source of radiation of these sensors (Zhou et al. 2015).

Halogen lamps have now been replaced by LED. Most sensors use LED plus an integrating sphere. The effect of scattering of solar radiation or absorption of radiation can be solved with a diffuser plate. The use of this plate provides a good Lambertian characteristic to the system, and PTFE (Teflon) is the main coating material used for the diffuser plates. Calibration based on solar radiation for sensors in orbit, in the visible and near-infrared bands, is currently the main method used (Zhou et al. 2015).

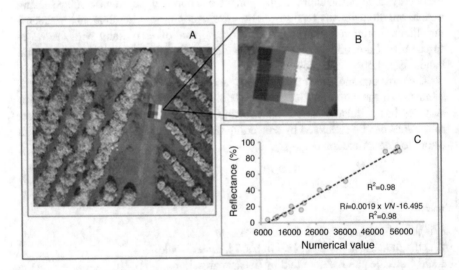

Fig. 4.6 Representation of the empirical line method of radiometric calibration: (**a**) aerial image corresponding to the NIR spectral band (820–860 nm) of the RedEdge-MX sensor in a coffee plantation, (**b**) targets with known reflectance, and (**c**) linear equation for the correction of numerical values for reflectance

4.4   Choice of Remote Sensing Platform

It is important to know that, for each type of application, the sensor to be chosen for data collection should have spatial, radiometric, spectral, and temporal resolutions compatible with the study objective. Spatial resolution refers to the size of the smallest element in an image, corresponding to the size of the smallest object that can be identified in that image. Radiometric resolution refers to the smallest variation in intensity possible to be detected by the sensor, represented by the number of levels of the numerical values of the pixels, as explained earlier. Each type of target shows a specific behavior in relation to its interaction with the EMR, which gives each type of object its spectral signature. Spectral resolution refers to the number of bands and the width of the spectral range of each band. Thus, a sensor with a high number of spectral bands with narrow wavelength ranges is a sensor of high spectral resolution. Finally, the temporal resolution concerns the time it takes for the same target to be revisited by the sensor.

The choice of platform will depend on the intended purposes, and there is no rule for this definition. Orbital platforms have the potential to acquire information in large areas, but they tend to lose spatial and temporal resolution. In addition, the spectral data acquired are greatly influenced by atmospheric conditions. Aerial platforms can also obtain data from large areas; however, with the gain in terms of spatial and temporal resolution, they are the most used to characterize spatial and temporal variability at the field level. Terrestrial platforms are used at the local level, often at specific points, requiring more efforts to acquire spectral data. These platforms can also be used for variable-rate application of inputs, both in real time and through maps.

Generally, the collected data are used to determine spectral vegetation indices, which can be used to characterize agricultural vegetation. From the vegetation indices, in digital agriculture, it is possible to identify spatial variability of plants, estimate yield, evaluate the nutritional deficiency of plants, and identify pests and diseases that affect a crop field.

4.5   Applications

Remote sensing is a very important tool for managing farms and plantations that adopt digital agriculture techniques. We can list—as some of the main applications—the use of sensors to indicate nitrogen deficiency in plants. One of these sensors is the GreenSeeker®, an optical sensor that calculates the NDVI of plants to estimate their nitrogen deficiency and perform variable-rate application. WeedSeeker® and WEED-it® are optical sensors that perform localized herbicide spraying, spraying only weed-covered areas, hence saving product and promoting a less environmentally aggressive operation.

Images obtained by sensors onboard orbital platforms have become a milestone in the science of remote sensing. Several orbital platforms provide an image of the entire Earth's surface. There are almost 6000 satellites in orbit, but only about 40% of these are in operation. Most satellites in orbit are used for communication purposes. A little over 400 satellites are used for Earth observation and nearly 100 for navigation/GPS purposes (Wood 2020). The United States, China, and Russia dominate the list of countries with satellites in operation, with about 1897, 412, and 176 satellites, respectively (UCS 2021).

Some platforms are presented in Table 4.1. Among the platforms presented, the LANDSAT satellite family coordinated by the US National Aeronautics and Space Administration (NASA) has been providing images of the Earth's surface for free since 1972, which has made this platform the most used in remote sensing studies. LANDSAT family's last satellite, Landsat 8, entered orbit in 2013. This satellite has two sensors onboard: the Operational Land Imager (OLI), which records images in the visible and near infrared region, with spatial resolution of 30 m, and the Thermal Infrared Sensor (TIRS), which records images in the thermal infrared region, with spatial resolution of 100 m. For years, the LANDSAT project has provided public domain images with the best spatial resolution; however, in 2015, the European Space Agency put into orbit the Sentinel 2 satellite, which carries the Multispectral Instrument (MSI) sensor onboard and has since been imaging the Earth's surface with spatial and temporal resolutions in the visible and near-infrared region better than those of LANDSAT (Table 4.1).

In agriculture, orbital images have had several applications, such as studies focused on land use and occupation, digital soil mapping, study of the nutritional status of crop canopies, and biomass estimates in closed canopy crops such as

Table 4.1 Spatial, radiometric, temporal, and spectral resolution of some satellites in operation

Satellite	Spatial resolution (m)	Radiometric resolution (bits)	Temporal resolution (days)	Spectral resolution
WORLD VIEW 3	1.24/0.30/3.70	11	<1	Multispec/pan/SWIR
GEOEYE	2/0.50	11	3	Multispec/pan
SPOT 6/7	6/1.5	12	Daily	Multispec/pan
RAPIDEYE	6.50	12	5.5	Multispec
TH-01	10/5 and 2	8	5	Multispec/pan
SENTINEL 2	10. 20 and 60	12	5	Multispec
LANDSAT 8	30/15/100	16 or 8	16	Multispec/pan/thermal
CBERS-04A (WFI/MUX/WPM)	55/16.5 and 8/2	5/31/31	10/8/10	Multispec/multispec/multispec and pan
IKONOS	4/1	11	3	Multispec/pan
QUICK BIRD	2.4/0.60	11	2.4	Multispec/pan

Pan panchromatic, *SWIR* shortwave infrared, *Multispec* multispectral

fodder, and in the context of rural property management, orbital images have been assisting in the definition of management zones. However, despite the evident applicability of orbital images in the agricultural environment, some of its characteristics, such as low spatial and temporal resolution, and the presence of clouds in the recorded scenes limit their use on some occasions. Some satellites offer images with better resolutions, such as WorldView-3, which provides spectral bands with spatial resolution of less than 1 m and temporal resolution of less than 1 day; however, the scenes have high costs, which often makes their acquisition unfeasible.

In studies that require greater spatial and temporal resolutions, images obtained by aerial platforms have been very useful. With the advent of unmanned aerial vehicles (UAVs), the cost for acquiring high-resolution multispectral images has been considerably reduced. This reduction in costs made the use of UAVs very attractive to the agricultural segment. The emergence of UAVs, together with the technological advance in the production of multispectral sensors with increasingly affordable costs, has been consolidating their use for remote sensing purposes in agriculture. The high spatial resolution of the images obtained by sensors onboard UAVs enables the acquisition of information with a high level of detail of crops, which broadens the horizon of the application of remote sensing as a tool for digital agriculture. In addition to the high spatial resolution, UAVs have also become an ally in temporal studies, since temporal resolution can be adjusted to the problem. Several remote sensors can be coupled to UAVs, including thermal sensors (emitted infrared), multispectral sensors (up to two dozen bands), and even hyperspectral sensors (hundreds of bands).

Some manufacturers (Table 4.2) already offer specific cameras for use in UAVs with application in agriculture. There are several options of sensors on the market; however, the higher their spectral resolution, the higher their cost. The sensors that have the most affordable costs are modified sensors, for example, the Survey3, manufactured by MAPIR®. This sensor is originally a color camera that has been altered to record in one of its channels bands in the near-infrared region; despite its lower cost, the reliability of the spectral information obtained has still been discussed. The RedEdge sensor, manufactured by MicaSense®, is a multispectral camera that

Table 4.2 Some sensors for use in UAVs available on the market

Camera	Number of spectral bands	Thermal band	Spectral resolution	Manufacturer
RedEdge-MX	5	0	Multispec	MicaSense
Altum	5	1	Multispec	MicaSense
Survey3	3	0	Multispec	MAPIR
Kernel	6	0	Multispec	MAPIR
MCAW	6/12	1 or 2	Multispec	Tetracam
Parrot SEQUOIA	4	0	Multispec	Airinov
PIKA L	281	0	Hyperspec	Resonon
HSC-2	1000	0	Hyperspec	MOSAICMILL

Multispec multispectral, *Hyperspec* hyperspectral

simultaneously captures five independent bands with narrow width, namely, the RGB bands, the NIR band, and the RedEdge. This last one mentioned is located at the threshold between the visible spectrum and the near-infrared, making it possible to investigate physiological characteristics of plants, which are not observed with vegetation indices that use visible bands, like NDVI, because these indices often tend to have their values saturated, not being sensitive to physiological changes.

Some more technological sensors already record one of their bands in the thermal infrared region, for example, the Altum sensor, manufactured by MicaSense®, and the MCAW sensor, manufactured by Tetracam®, which can be optionally equipped with one or two thermal cameras. However, these sensors have higher costs. The manufacturer FLIR® has several models of thermographic cameras that can be used to determine plant and soil temperatures, assisting in the detection of crop water stress. In addition, it makes it possible to evaluate the variability of plant water status in space and time. Some examples of camera models from this manufacturer used for water stress detection are E5, model I50, model T640, and model A310 (for studies with image acquisition through UAVs), among others.

Given the characteristics of UAV-assisted remote sensing presented above, there are numerous applications of this tool in digital agriculture. The high spatial resolution of the images obtained by UAVs are assisting practices such as high yield phenotyping for genetic improvement of crops, detection of diseased plants, detection of pest-infested plants, inferences on the status of nutrients in plant tissues, detection and counting of planting failures, among other variabilities found within the cultivated plots. The fact that the temporal resolution of UAVs can be adjusted to each situation allows its use with high efficiency in temporal monitoring, for example, in determination of the harvest point in crops such as soybean and coffee, study of the temporal behavior of epidemics that affect agricultural crops, estimates of crop yield, and determination of the appropriate time for agricultural management practices such as irrigation, fertilization, and weed control.

Abbreviations/Definitions

- Electromagnetic radiation (EMR): It is a junction of a magnetic field with an electric field that propagates carrying energy. Light is an example of electromagnetic radiation.
- Spectral signature: The characteristic of each target on the Earth's surface, regarding its response to EMR, is called a spectral signature. It corresponds to the reflectance curve across the electromagnetic spectrum.
- Vegetation indices: Combinations of two or more reflectances at specific wavelengths or spectral bands to generate an index that is related to a vegetation characteristic.
- NDVI: Normalized Difference Vegetation Index, based of red and infrared bands, is one of the most popular indices.
- Digital images: Information recorded by digital cameras and stored in numerical matrices. Digital images are generated by a process called digitization. The digitization process makes possible an analog signal, from the energy that reaches the digitizer sensor, to be transformed into digital information.

- CCD and CMOS: Charge-coupled device and complementary metal-oxide-semiconductor are semiconductor sensors for capturing images formed by an integrated circuit that contains a matrix of coupled capacitors. Under the control of an external circuit, each capacitor can transfer its electrical charge to another neighboring capacitor.
- Pixel: Each element of the matrix that forms an image is called a pixel, which is presented on a scale of integers greater than or equal to zero.
- Spatial resolution: It refers to the size of the smallest object that can be identified in an image.
- Radiometric resolution: It refers to the smallest possible intensity variation to be detected by the sensor, represented by the number of levels of the numerical values of the pixels.

Take Home Message/Key Points

- Remote sensing can be an important tool for implementing digital agriculture.
- Information obtained by optical sensors is helping in the management of farms because the acquisition of information can be done in a short time and with a reasonable cost.
- With the technological advances, increasingly affordable optical sensors that can be transported by UAVs are becoming available on the market. In addition, these sensors allow images to be obtained with more and more details of the fields.

References

Formaggio AR, Sanches ID (2017) Sensoriamento Remoto em agricultura. Oficina de Textos, São Paulo

Lebourgeois V, Bégué A, Labbé S, Mallavan B, Prévot L, Roux B (2008) Can commercial digital cameras be used as multispectral sensors? A crop monitoring test. Sensors 8(11):7300–7322

Liang S, Wang J (2019) Advanced remote sensing: terrestrial information extraction and applications. Academic

Rouse JW, Haas RH, Schell JA, Deering DW (1973) Monitoring vegetation systems in the Great Plains with ERTS. In: Proceedings of 3rd earth resources technology satellite-1 symposium, 10–14 December, vol 1. NASA, Washington, DC, pp 309–317

UCS. Union of Concerned Scientists (2021) UCS satellite database. Available at: https://www.ucsusa.org/resources/satellite-database. Accessed in Feb 2021

Wang T, Thomasson JA (2019) Plant-by-plant level classifications of cotton root rot by UAV remote sensing. In: Autonomous air and ground sensing systems for agricultural optimization and phenotyping IV. International Society for Optics and Photonics, p 110080N

Wood T (2020) Who owns our orbit: just how many satellites are there in space? Available at: https://www.weforum.org/agenda/2020/10/visualizing-easrth-satellites-sapce-spacex/. Accessed in Feb 2021

Zhou G, Li C, Yue T, Jiang L, Liu N, Sun Y, Li M (2015) An overview of in-orbit radiometric calibration of typical satellite sensors. The international archives of the photogrammetry, remote sensing and spatial information sciences. In: International workshop on image and data fusion. v. XL-7/W4. https://doi.org/10.5194/isprsarchives-XL-7-W4-235-2015

Chapter 5
Geoprocessing Applied to Crop Management

Lucas Rios do Amaral and Gleyce Kelly Dantas Araújo Figueiredo

5.1 Introduction

With the speed of georeferenced data that are made available, either using yield monitors or image sensors, it is essential that these data are processed quickly and reliably, allowing decision-making. Thus, in the era of digital agriculture, an environment that is capable of handling and managing geospatial data and providing quantitative and qualitative information is mandatory.

Geoprocessing is fully capable of dealing with geographic information, and within this universe, geographic information systems (GIS) are capable of performing complex analyses based on the integration of different geospatial components. To this end, it is necessary that users know the data formats used, ensuring the quality of data obtained from different sources and adopting the appropriate analyses for each situation. Only then will the final product be reliable and able to guide decision-making processes.

In this chapter, we present in the first part the components of a GIS, the ways of obtaining data, and the basic analyses available in this environment. In the second part, we discuss some digital image processing techniques. In the last part, we present ways of integrating data, addressing the concept of management zones, and the integration of spectral and meteorological data.

L. R. do Amaral (✉) · G. K. D. A. Figueiredo
School of Agricultural Engineering, University of Campinas (FEAGRI/UNICAMP),
Campinas, São Paulo, Brazil
e-mail: lucas.amaral@feagri.unicamp.br; gleyce@unicamp.br

D. Marçal de Queiroz et al. (eds.), *Digital Agriculture*,
https://doi.org/10.1007/978-3-031-14533-9_5

59

5.2   Geographic Information Systems (GIS)

Geographic information systems, otherwise known as GIS, are based on the integration of different components to analyze and interpret geospatial data, that is, data for which location (coordinates) is an important attribute to be considered. They consist of a software application with characteristics that are specific to this type of system and a hardware suitable for a proper operation with such type of software, a user who operates such software and is able to apply the appropriate analysis techniques for the desired purpose (also called as analyst or interpreter), and the data itself, which can have several characteristics, the fundamental one being the availability of information about their spatial location, usually represented by spatial coordinates (latitude and longitude). Thus, GIS can be defined as a set of functions with advanced capabilities for storing, accessing, manipulating, and visualizing georeferenced information (Burrough 1986).

A recurring mistake is to assign the title of GIS to the analysis software. However, a single component does not represent a system, since it is not capable of generating a product by itself. In this sense, one way to differentiate the software that makes up this system from other simpler ones, which do not have the capacity to perform a wide range of geospatial analyses, is to call it *GIS software*. GIS software must have some specific features that allow for the computational treatment of georeferenced data to generate products such as maps, tables, and graphs. These essential tools include the visualization and manipulation of geospatial data, such as feature edition, cartographic redesign, spatial query, data format conversion, map algebra, plotting of results, data storage and recovery, and editing and alteration of databases, among others.

ArcGIS and QGIS are widely known GIS software. *ArcGIS* is a popular and complete commercial software by ESRI available in the world market. *QGIS*, previously called Quantum GIS, is an international free and open-source software initiative that is in constant development and has many features that are equivalent to those of ArcGIS (to learn more about the initiative, see www.qgis.org/). Other available software includes, for example, SAGA GIS, SPRING, GRASS, and gvSIG, which are, however, initiatives with a reduced number of users and less availability of support material on the Internet (tutorials and discussion forums) compared to the first two. Moreover, more and more programming packages for this purpose to be used in environments such as R and Python have been emerging.

(a) *Data Model*

The procedure for opening a data file in GIS is called "import," and each data set that is being worked on is called "information layer." These data can be divided into two types of models: vector and raster models.

5.2.1 Vector

This data format is adopted when the user wants to represent features with well-defined limits (Fig. 5.1) by working with points, lines, and polygons, considering that each layer of information must represent only one of these three. Therefore, in agriculture, points can be used, for example, to represent the places of soil samples or the presence of obstacles for agricultural machines; lines can represent the drainage networks and the path a machine must follow; and polygons can represent any feature with closed lines, such as field boundaries and dam area. The vector format is simple to encode, taking up little disk space, and is most suitable when the shape of the features is important to the analysis. The most commonly used file format for vector data is the shapefile format, which is composed of the file with the .shp extension but is also always linked to at least three other files (.shx, .dbf, and .prj). Therefore, it can be composed of up to six files with different extensions, which must always be together and have the same name to successfully import this type of data.

5.2.2 Raster

This data format is adopted when the user wants to represent a continuous variable that is present throughout the study area (Fig. 5.1), functioning as if it were an image of the entire area. This file is basically composed of equally sized pixels over

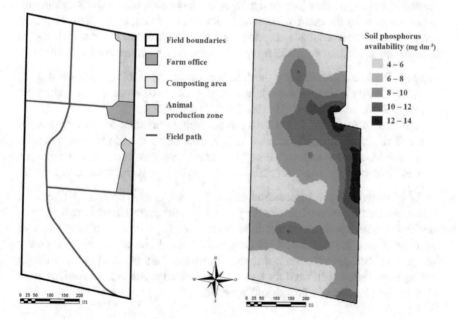

Fig. 5.1 Representation of data in the vector (left) and raster (right) formats. (Source: The authors of this chapter)

its length, as observed in digital photographs when zooming in enough to see the "squares" making up the image. In agriculture, this type of file is the product of satellite and drone imaging, used to obtain an image of the target (e.g., soybean field). It also comprises maps interpolated from data in the vector format, with the aim of, for example, representing the spatial variability of a certain soil property (Fig. 5.1). The raster format is ideal for representing attributes with high variability throughout the study area and for cases where it is necessary to perform some kind of mathematical operation between information layers, since all of these layers will be linked to the coordinates of the same pixels. The file format that is most commonly adopted for vector data is GeoTIFF, but several other common image formats can be used, such as .jpeg and .bmp.

(b) *Creation of Databases*

Data are fundamental components of a GIS, and their quality is essential so that the products generated represent the reality of the study field and can be confidently used in decision-making processes. It is always important to remember the maxim "trash in, trash out." There are basically two ways of obtaining the data needed for the study: office surveys and field campaigns.

5.2.3 Data acquisition

With the "digitalization of agriculture" and the advances in the development of sensing technologies, data storage methods, and more elaborate analysis techniques, inferences and decisions are often made without ever going to the field. There are several data sources available on public and private platforms that can provide the information needed for a particular type of study. Two examples can be highlighted:

– Earth Explorer, provided by the US Department of Geology (US Geological Survey: earthexplorer.usgs.gov), is a source with even more varied options for image databases obtained via orbital remote sensing.
– Agricultural machines themselves are becoming important sources of data for use in decision making processes. This is because more and more technologies are developed every day, with sensors measuring various parameters of machines while they operate, and even transferring them to processing units in real time.

All of these data can be accessed on the Internet, cutting back on the need for field evaluations. However, despite the potential for the performance of remote analysis and various applications that have been emerging with the "digitalization of agriculture" and the so-called Agriculture 4.0, it should be noted that, to properly interpret the data, as well as refine the algorithms and standards used by the Artificial Intelligence, there is still need for agronomic knowledge and field campaigns, with an effective adoption of a "hands-on" approach.

5.2.4 Field Data Collection

In this case, field campaigns may have different purposes, depending on the nature of the study. Thus, the field data can be the main data of the study, complementary to the data already collected in the office, or they may be used to validate the inferences remotely made.

- In the first situation, some types of analysis may have been previously performed in the office, but the investigation is not complete without the main data that will be collected in the field. An example of this type of survey is grid sampling to map the soil's fertility. In this technique, GIS are used to generate a sample grid, either based on the field boundary and regularly spaced sampling points or on more advanced analyses, such as the integration of satellite imagery, digital terrain modeling, and proximal sensor mapping (e.g., soil apparent electrical conductivity sensing) to indicate the most representative sampling points.
- In the second situation, data that will complement the information obtained using GIS can be collected. An example would be collecting leaf samples to, based on their nutrient content, refine a fertilizer recommendation algorithm, developed using historical productivity data, satellite images, and the farmer's own knowledge.
- Finally, the field data may be collected to validate the information obtained using GIS. A classic example is to check which pest is damaging the crop canopy. This damage (symptom) is identified based on a satellite imagery time series or even by drones, making it mandatory to know which species is present in that field spot in order to properly implement the site-specific management.

5.3 Data Analysis in GIS

Some types of analyses and/or tools are typical of GIS, especially when assessing agricultural data. In this book, several of these analyses are covered, but we present below some very simple functionalities that are almost (or ideally) always performed when working with georeferenced data.

5.3.1 Obtaining and Converting Text Files

This type of file format is common, since it allows importing electronic spreadsheets (e.g., Excel) into GIS software. We often manipulate data sets in spreadsheets, such as to combine a point file coordinates (vector) with the soil analysis results or to assess them in whatever software the user feels more comfortable and/ or that offer additional resources, but which must be subsequently spatially viewed in GIS. In these cases, the data coordinates (latitude and longitude) will comprise two of the variables (columns) in the spreadsheet. These spreadsheets must be saved in simple formats, usually in .csv (comma-separated values) or .txt (tab-delimited text files), so they can then be imported into a GIS software. In the import process, the analyst must indicate the columns of the coordinates and the projection system (datum) used to survey the data, because, unlike the ShapeFile and GeoTIFF files mentioned above, such simpler file format does not contain integrated cartographic information. As it is a view-only format, once imported, it usually needs to be converted to another format (usually Shapefile), so that it becomes manipulable within the GIS software. In this way, the text file is considered as data input, since vector-type data are generated from it.

5.3.2 Spatial Query

It is a basic GIS operation in which the data set used can contain several attributes. In addition to the geographical coordinates, it is often sought to select features (or points) that are within a given value (or distribution) limit or that contain a certain characteristic (e.g., chemical properties of soil samples). In this way, the query (or selection) can be performed based on the value of the attribute under analysis. Other types of spatial query make use of the feature's own spatial location, or even of the combination of attributes available in the information layers, in which case the structured data language (SQL) can be used.

5.3.3 Exploratory Data Analysis

Although this type of analysis is not specific to GIS, it is a good assessment practice regardless of the data source. This is because the quality of the information generated depends on the quality of the input data, and the greater its reliability, the greater the certainty with which decisions based on the results of the GIS analysis can be made. In this sense, descriptive statistics are commonly used to summarize and describe the data set by calculating the measures of central tendency (e.g., mean and median) and variability (e.g., standard deviation, variation coefficient, maximum and minimum values, asymmetry, and kurtosis). These measures can also be represented graphically, which can help the analyst make more intuitive interpretations; in this sense, the most used products are histogram and boxplot (Fig. 5.2).

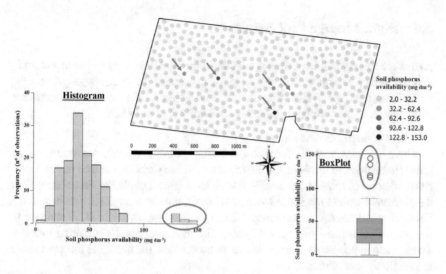

Fig. 5.2 A field where soil sampling was performed and the phosphorus content available in the soil (P) was quantified. The red arrows indicate potential outliers identified by the histogram and the boxplot of P values (red ellipses). (Source: The authors of this chapter)

5.3.4 Data Filtering

A usual procedure when analyzing data from sampling or sensors embedded in tractors or any other platform (orbital, aerial, drone, terrestrial) is the data set filtering. This practice has the purpose of removing observations (points/values) deemed doubtful in comparison with their neighborhood's behavior, that is, information that does not follow the same pattern as the rest of the data (outliers). It is often identified using statistical criteria, such as by removing all samples that exceed the mean value, more or less three times the standard deviation, or 1.5 times the inter-quartile range. However, when dealing with geospatial data, this type of statistical procedure should only indicate *possible outliers*. Therefore, the analyst must evaluate the spatial distribution of the points (Fig. 5.2) in order to understand whether the values considered as outliers really differ from the standard of a region (i.e., do not behave like neighboring samples), in which case they must be discarded, as they may have been caused by sampling or measurement errors. Thus, when it is assessed that a data point may not represent the region it should be representing, it should preferably be excluded from the analysis to make the decision-making process more reliable.

5.4 Digital Image Processing

A digital image (satellite or drone) contains important information, but to extract it, it is necessary to apply different types of processing (digital image processing, DIP). Thus, the ways of acquiring images as well as correction techniques, enhancement, and even interpretation elements are briefly discussed here.

(a) *Ways of Acquiring the Images*

There are numerous platforms for acquiring satellite images, depending on the sensor to be used, in terms of processing them. These platforms can be freely accessible, where free images are usually found. Examples include the image catalog of the National Institute for Space Research (INPE), the Earth Explorer platform of the US Geological Survey, or the Google Earth Engine platform, which, in addition to images, offers a great diversity of geospatial data and cloud computing process. There are also private platforms, where, in most cases, the available images belong to satellites for commercial use.

Another way to acquire satellite images is with software applications that directly connect users to the image database. It is possible to download some satellite images using an add-on in the QGIS (Semiautomatic Classification) software, or packages in R or Python language. In both examples, users must enter the access name and password for the database containing the images they wish to obtain.

Once the user has defined the platform, it is necessary to know the level of processing applied to the available images.

It is important to note that the acquisition of images by drones is completely different from what has been described here. In this case, it is necessary to create a flight plan (height and direction of flight, area to be imaged, percentage of overlap between images, etc.), always considering the type of sensor (camera) that will be onboard to the drone.

(b) *Processing Level*

Remote sensing images in their raw format usually have flaws or some kind of deficiency. Thus, to use them, it is necessary to apply the appropriate corrections. The processing level will change depending on the aim of the study. For example, when acquiring an image with high spatial resolution from an agricultural area to check for the presence of weeds and site-specific spraying, its geometric distortion, if not corrected, will cause large-scale location errors.

5.4.1 Geometric Correction

The acquisition of images at the orbital (satellites) or aerial (planes or drones) level is always subject to spatial distortions, whether natural (the Earth's rotation and relief) or related to the platform carrying the sensor (variations in the platform's altitude, speed, and rotation axes). Thus, these images do not have a correct correspondence regarding the positioning of the object or phenomenon on the Earth's surface, which makes it impossible to accurately perform GIS analyses, for example.

Georeferencing is a geometric transformation that corrects distortions caused by the Earth's rotation and the platform, relating the coordinates of an image to the coordinates of a previously georeferenced map or image. This process makes use of control points that can be easily identified in the image, which are applied using polynomial equations. The orthorectification process, on the other hand, corrects the distortions caused by the elevation differences over the terrestrial surface using a Digital Elevation Model (DEM) (altimetric representation of the terrain, disregarding, e.g., buildings and trees in the area), in addition to the orbital parameters depicting sensor's altitude in relation to the Earth's surface at the moment the image was taken. Moreover, it is also possible to use terrestrial control points to obtain accurate coordinates.

Generally, many agencies correct these distortions before making the images available. For example, images obtained by LANDSAT or the Chinese-Brazilian CBERS satellite (China–Brazil Earth Resources Satellite Program) have different processing levels.

5.4.2 Atmospheric Correction

In a satellite image, each pixel is assigned a digital number (DN), which corresponds to the intensity of electromagnetic energy reflected by a specific target and measured by the sensor (radiometric resolution). When the objective is to obtain the biophysical characteristics of a target, it is necessary to convert DN to radiance at the sensor level or at the top of the atmosphere and then to surface reflectance values.

Among the available atmospheric corrections model, we can mention the Dark Object Subtraction (DOS) method proposed by Chavez (1988). However, one of its major criticisms is that it does not consider the absorption effects of atmospheric gases, taking into account only scattering phenomena.

Other options include radioactive transfer models that consider not only the sensor's characteristics but also the environmental conditions in which the image was acquired. This atmospheric correction is frequently performed using the Second Simulation of a Satellite Signal in the Solar Spectrum, also known as 6S. In general, these atmospheric corrections are easily implemented in software applications; for example, in QGIS, it is possible to apply the DOS method using the *Semi-Automatic Classification* plugin or even the image download packages in R language themselves.

Currently, many sources already provide data with a high processing level. The Earth Explorer platform, for example, offers Landsat 8 OLI (Operational Land Imager) images corrected for surface reflectance. Another example is the Google Earth Engine platform, which allows using both data corrected for top of

atmosphere reflectance (TOA) and surface reflectance (SR) or bottom of atmosphere reflectance (BOA), from several satellites.

(a) *Image Enhancement Techniques*

The contrast of digital images captured by sensors is perceived by the human
visual system as low quality, which can be improved using enhancement techniques;
among these, histogram image manipulating is one of the most common. Remote
sensing images are acquired at certain radiometric resolutions, and their gray levels,
ranging from white to black, may be too dark or too light. In both cases, the visual
interpretation process would be easier if the contrast were applied to these images'
histogram. Generally, in a histogram, the DN values do not spread across the gray
level range, which can make its visual interpretation difficult. Thus, increasing or
decreasing the contrast of the histogram is often done subsequently that an interpreter can extract as much information from an image as possible. Regardless of the
work's purpose, it is necessary to know the portion of the histogram to be highlighted, since each type of contrast has a different function, leading to different
results.

It is important to emphasize that the purpose of image enhancement is improving
the quality of information that is already contained in the image, not being capable
of providing new information.

(b) *Basic Visual Interpretation Elements*

Visual interpretation is performed to obtain information about the Earth's surface
based on the spectral response of the targets and on some aspects of the image itself,
as follows:

- *Tonality*: Related to the reflectance properties of the objects and the spectral
band, the interpreter is analyzing. Each target on the Earth's surface absorbs and
reflects light differently and at different wavelengths. For example, when analyzing a grayscale image in a region occupied by a forest and if that image refers to
the near-infrared spectral band, the tonality of the target will be close to white;
on the other hand, if that same image refers to the red spectral band, the tonality
will be dark gray. That is because the green vegetation reflects a lot of energy in
the near-infrared spectral band and has high energy absorption in the red band.
- *Texture*: This component is related to the shadow and tonality frequency of targets. When analyzing an image with three different targets, such as natural forest,
agricultural cultivation area, and a water body, the forest will have a rough texture due to the different vegetation strata (tree sizes), as well as the shade caused
by this height difference. On the other hand, the texture of an agricultural cultivation area, for example, a sugarcane or soybean field, will be smooth to rough,
since the canopy of the plants should be relatively homogeneous after canopy
closure as they are planted at the same time and are thus at the same stage of
development. Lastly, in rivers and lakes, the texture is smooth, since the target's
variations in height and tone are not significant.
- *Shape*: The shape is associated with what the area is used for and can be divided
into regular (with human intervention) and irregular (without human intervention).

For example, plots with agricultural crops or forests may have a rectangular shape, and areas irrigated by a central pivot have a circular shape, while natural areas, such as a natural forests or rivers, have irregular shapes.

- *Pattern*: It is associated with the spatial arrangement of objects in the landscape, that is, the way in which the distribution of objects is repeated within the area, which could be random, systematic, linear, and interwoven, among others. For example, coffee is planted in rows, thus linearly and systematically.
- *Shading*: Most images are obtained by the satellites about 2 h before or after noon, to avoid excessive shading. However, shading can help identify an object. For example, fences dividing agricultural areas can after only be identified by their shading; in addition, when identifying mixed cropping areas, due to the differences in plant height, the shading will result in different textures.
- *Size*: Considered one of the most distinctive features. If objects are visible in an image, it is possible to determine their size in the scene by comparing their dimensions.
- *Context or situation*: Objects can be identified based on the location associated with a certain phenomenon or activity. For example, a building in the middle of a pasture will not be interpreted as a machine shed, but as a livestock shelter.
- *Color*: When working with target interpretation, color is one of the interpreter's greatest allies. By combining three spectral bands, it is possible to extract more information from the study image, as it allows introducing not only color (hue) but also saturation (gray tones), thus providing greater richness of details in different bands of the spectrum.

(c) *Color Composites*

The human visual system is able to distinguish approximately 30 shades of gray; however, when working with colors, that number changes to thousands. Thus, working with grayscale remote sensing images is a difficult task, as its interpretation is limited to this human capacity.

In the visual interpretation process, when working with multispectral sensors (multiple bands), it is possible to create a single image with three different bands of the electromagnetic spectrum to extract as much information as possible based on the colors. This process is possible due to the RGB space and the basic color theory (further discussed by Crósta (2002) and Moreira (2011)), proposed by Thomas Young who, when projecting light in superimposed red, green, and blue circular filters, verified two color formation processes: the primary additive process, which uses a mixture of two or more primary colors (red, green, or blue) to obtain secondary colors (magenta, yellow, or cyan), and the primary subtractive process, which subtracts the primary colors to produce secondary colors. It was also observed that overlapping the primary additive colors resulted in white and overlapping the primary subtractive colors resulted in black. Moreover, the variation in the intensity of the additive primary colors can result in a wide range of colors and shades.

The RGB space allows working with the so-called color composites, in which a synthetic image is created by combining three spectral bands. This is because the remote sensing images are expressed in gray levels and, as explained previously, each pixel has a variation in the intensity of the gray level, ranging from black to white. Thus, knowing the RGB filters and the spectral behavior of each target allows deciding which spectral bands can be used to visualize a target in a given color.

When interpreters wish to analyze an area with different agricultural uses, it is important that they are aware of the spectral curve of vegetation, soil, and water. Figure 5.3 is an RGB-564 color composite of Landsat 8 OLI, where there is cultivation of sugarcane and annual crops (soy and corn), riparian forest, bare soil, and a river. When composing this synthetic image in the RGB space, the following colors are obtained:

- Filter R (red) was applied to band 5 (near-infrared). In this spectral band, the vegetation shows high reflectance, and consequently, light gray intensities will be observed in the image. Thus, the vegetation, in general, will be red in color.
- Filter B (blue) was applied to band 4 (red). In this region of the spectrum, the soil's reflectance is relatively high, and the shades of gray in the image are lighter, resulting in shades of blue for the soil in the synthetic image.
- Filter G (green) was applied to band 6 (SWIR). In this spectral band, the soil has high reflectance; however, as there are two bands with a predominance of lighter soil tones in this image, there will be a mixture of colors between green and blue, resulting in soils with blue and/or cyan tones, depending mainly on the type of soil and its water content.
- In this RGB-564 color composite, the vegetation is not entirely red, since the shortwave-infrared spectral band was also used and, in this band, the reflectance is dominated by the water content in the leaves. Thus, it is possible to distinguish plants with higher water content, such as natural or planted forests and sugarcane, which will be red in color, from annual crops with lower water content, such as soybeans and corn, which will appear yellow or orange, depending on the color mix between red and green.

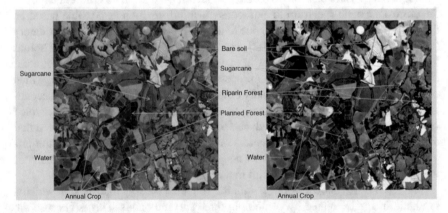

Fig. 5.3 RGB564 color composite, false color (left), and RGB432 color composite, natural color (right). (Source: The authors of this chapter)

- Due to the high absorption of electromagnetic radiation from water, water bodies will be black in color. However, is there is a presence of sediments in water bodies, it will appear closer to blue.

This type of color composite is called a false color composite, that is, composites made with other bands of the electromagnetic spectrum in addition to the visible spectrum, allowing the user to characterize the target, for example, the type of vegetation. It is also possible to create a natural color composite, made with spectral bands of the visible spectrum by applying the RGB filter to bands of the red, green, and blue region, which results in the synthetic image human eyes see in the real world (Fig. 5.3).

In general, there is no right or wrong color composite but rather, composites that are commonly used to characterize some targets on the Earth's surface, highlighting certain features of the targets to facilitate the visual image interpretation. What users should take into consideration is which color composite will allow them to extract as much information as possible from an image.

(d) *Image Classification*

Remote sensing image classification consists in associating each pixel of the image with a label (e.g., land use classification), which begins with the recognition of targets based on the basic visual interpretation elements described above. The second step is to label these previously recognized objects, generating themes or classes. This results in a thematic map with the spatial distribution of each element recognized in the image.

To understand how image classification works, it is essential to understand the attribute space graph of the image being classified. Attribute space refers to the frequency of distribution of gray level intensities of two or more spectral bands. When working with only one spectral band, two or more targets can often have similar gray level intensities, which makes it difficult to separate them. By adding two or more bands, we will observe greater separation between the targets depending on the spectral response of each of them in the different bands. This is possible to verify based on attribute space. In order to correctly identify the objects in the image, it is important to consider a group of pixels with similar behavior, so as to verify whether this is an expected variation of the target and not only one pixel.

In Fig. 5.4, it is intended to separate targets A and B. When observing the gray level distributions in only one band, these two targets overlap; by adding the second band, it becomes possible to separate them. As the number of bands increases, the separation power also increases. Therefore, it is very common for some users to apply all possible bands of the sensor in the classification process.

There are two types of approaches to image classification: (1) unsupervised classification, when the user does not have enough information about this area or intends to carry out an exploratory analysis and (2) supervised classification, when the user has some knowledge about the area due to the availability of training data (labels).

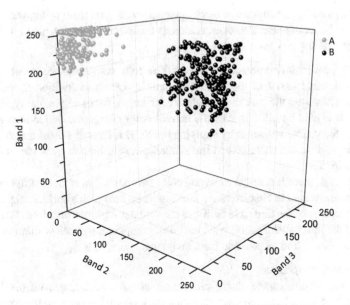

Fig. 5.4 Separation between similar targets based on attribute space. (Source: The authors of this chapter)

5.4.3   Unsupervised Classification

In this approach, it is the algorithm that defines which classes will be separated and which pixels will belong to those classes, based on statistical rules (mean, variance, standard deviation, etc.) and on the similarity between pixels. The grouping (or clustering) of pixels according to the spectral similarity between them is a very widespread approach when the user needs to perform unsupervised classification. In this case, only the number of classes that must be created is defined, and the algorithm distributes the pixels, according to their rules, throughout the study image.

5.4.4   Supervised Classification

In this approach, it is essential that users have prior knowledge of the area they intend to classify. To start this process, it is necessary to obtain representative samples of each target to be classified. Subsequently, these samples are used to train an algorithm based on statistical rules that will recognize similar patterns followed by the remaining pixels of the image. It is crucial to highlight that the stage of collection of training samples is important to obtain high-quality results, and the best option is going to the study location to register the occurrences of use and their geographical coordinates. Despite being the most suitable, this process usually takes time and has a cost. Alternatively, it is possible to visually inspect the image and assign a label to a set of pixels to then use these labeled sets as training samples. This is indicated when the user knows the study area well and can do this using an image with high spatial resolution.

Another important factor that must be considered is the representativeness of each sample, that is, the extent to which each sample represents the variability of each target in the image. Finally, to obtain good classification results, the user must divide these samples into three subgroups, namely, training, in which the behavior of each class is "taught" to the algorithm; testing, in which the model's ability to classify the area is evaluated based on training samples; and validation, in which the samples that were not considered in the previous steps are used to validate the final results (ability to predict unknown samples).

According to Zanotta et al. (2019), there are different supervised classification methods: those considered parametric, with a fixed number of classification parameters, and non-parametric classifiers, which learn functions with parameter numbers that vary according to the number of training samples. Examples of unsupervised and supervised classifiers can be found in the literature (Mather and Koch 2011; Zanotta et al. 2019).

5.5 Vegetation Indices

The extraction of crop biophysical parameters is essential to understand what is happening in the field. Thus, continuous monitoring of agricultural areas must be carried out to avoid loss of information over its development. One way to carry out the monitoring of vegetation is based on vegetation indices (VI), which allow monitoring the crop's temporal dynamics and vigor.

These indices are dimensionless radiometric measures that provide information related to the vegetation's activity and its interactions along the electromagnetic spectrum, such as leaf area index (LAI), percent vegetation cover, chlorophyll content, and biomass, among others (Jensen 2007). Thus, their use in agriculture is extremely important, because they allow monitoring the crop from the implantation stage (exposed soil) to the plant's maturation (loss of vegetative vigor).

There are countless vegetation indices available, and it is up to users to decide which one to consider depending on the information they need to extract. For example, in precision agriculture, it is common to use VIs to check the variability of biomass and relate it to soil fertility and crop productivity. In this case, a VI that uses the ratio of the red and near-red spectral bands is already sufficient to evaluate the variation in the canopy's vigor along the field. As the chlorophyll in the leaves tends to be related to the nitrogen absorbed by plants, when using a VI to estimate the amount of nitrogen, it is necessary to consider a spectral band where chlorophyll interacts with electromagnetic radiation; in this case, indices with the red-edge spectral band are often used. For applications such as stress verification, it is common to use VIs with the shortwave-infrared spectral band, in which the reflectance of vegetation is dominated by water content in the leaves.

Jensen (2007) provides a list with several VIs and their theoretical foundations, based on which it is possible to decide the best index for a given application.

5.6 Data Integration and GIS

When composing a geographic database, the analyst must keep in mind the objective to be achieved. This is important for two reasons: to know exactly what data will be needed and in what format they should be in as well as to ensure that the type of analysis chosen will provide the desired results. With such issues defined, it is possible to perform several types of data integration within the GIS.

5.6.1 Cross-Referencing Methods

Data integration from several data sources are frequently demanded for obtaining a certain product. Thus, great part of the data may not have the same spatial location, either because of the place where the sample was manually collected, the path followed by the machinery, or even as a result of the different imagery resolutions. This makes it difficult to compare information layers and either relate them in some way or cross-reference the data collected. Thus, to carry out studies with data from different sources, it is necessary to establish the same coordinate reference system (CRS) so that they can be spatially related. With this objective in mind, there are basically three ways of merging data: data interpolation, use of buffer extraction, or pixel resampling.

5.6.1.1 Data Interpolation

One of the most intuitive ways to compare data with different characteristics and sampling densities is to convert them, usually from the vector to the raster format, using a process called "interpolation." This procedure has the main objective of estimating the values of the variable under study on non-sampled locations to obtain data within the entire area of interest (Fig. 5.5). Thus, the analyst must often define the distance between each interpolation (this varies with the map's purpose), which will represent the pixel size of the raster file generated. With the interpolation of the different data being carried out for the same distance (or pixel size), all pixels will be in the same coordinate, allowing analyses (cross-referencing) to be performed between information layers. It is important to mention that there are several types of interpolators, with different complexity levels and purposes, resulting in slightly different products.

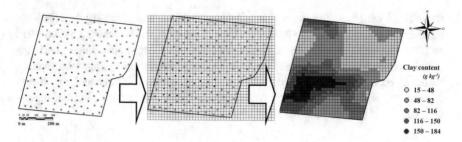

Fig. 5.5 Map with soil sampling points classified according to clay content (left); interpolation grid defined in GIS software (center); and continuous map, with interpolated data and pixel classification, according to clay content (right). (Source: The authors of this chapter)

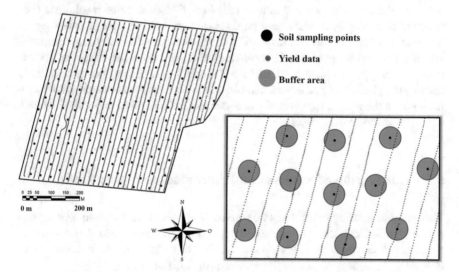

Fig. 5.6 Map showing the soil sampling points and data obtained using a crop yield monitor. The zoomed-in highlight depicts the region of influence generated by the buffer, showing that there will be at least seven yield data points for each soil sample, which will be used to calculate the mean yield value and relate it to the soil sampling coordinates. (Source: The authors of this chapter)

5.6.1.2 Use of Buffer Extraction

When it is not necessary to analyze the entire study area on a continuous basis and, consequently, perform data interpolation, an option is to use buffer data extraction, which is a typically tool found in GIS software. Such tool aims to relate data with different densities and locations and that are close to the same coordinate, within the limits (buffer) established by the analyst. This allows other types of data manipulation and analysis to be performed in GIS software or elsewhere, such as correlation or regression analyses. The buffer tool delimits areas of influence around one of the information layers (usually, the layer with the lowest sample density) containing data from the other layer (which are usually denser). Figure 5.6 illustrates an example in which, within the region of influence of the soil sampling data (buffer), there are at least seven crop yield data. Thus, the objective is to extract a crop yield value in the same coordinates of each of the soil sampling points. Such yield data may be either the closest point to the coordinates of the soil sampling coordinates a measure representing the yield points within an area of influence (e.g., mean, median or standard deviation), or even a value estimated using some kind of regression (trend surface, as proposed by Driemeier et al. (2016)).

5.6.1.3 Pixel Resampling

When the information layers are all in raster format, as in the case of studies based only on satellite imagery, these images may have different spatial resolutions (i.e., different pixel sizes), or there may be displacement between images obtained by different sensors. In this case, depending on the desired analysis, it may be necessary to ensure all images have the same resolution (i.e., same pixel size and coordinates). This type of operation is often called pixel resampling, and it consists in resizing the pixels using equations that consider their neighborhood and spatial relationships in the image. However, the analyst must keep in mind that some information may be lost in this process.

5.7 Integration of Spectral and Agrometeorological Data

The integration of agronomic data observed in the field with remote sensing and agrometeorological data is often carried out in order to understand the phenomenon occurring in the study area. It serves as a basis for technicians, farm managers, or governmental agency to make different types of decision.

There are several agrometeorological data that can be used to evaluate a crop in the field, such as temperature, precipitation, wind speed, and global radiation, among others. Additionally, there are several data sources, from meteorological stations, pluviometer, and remote sensors to mathematical models, which use all the previous sources to provide an adjusted value for the study variable. These sources differ in relation to their spatial and temporal resolutions. In the case of meteorological stations or pluviometer, the output data are specific, according to the geographic location of the equipment; in the case of remote sensors and atmospheric models, the data is obtained in a raster format (image), making it possible to spatialize the event; in all cases, the temporal resolution can vary from hours to days.

For farm-level scales, stations or pluviometer are sufficient; when the scale is regional (encompassing a large area), it is necessary to use data with the same spatial coverage, to verify the spatial variability of the event across the area. If it is not possible to use sources with spatial coverage, data interpolation from a network of meteorological station should be done, paying attention to the density of this network and to the spatial effects of events that occurred in the area.

Integrating agrometeorological and spectral data allows checking for areas affected by frost, for example, using a map of field, air temperature data, and profile of the crop development throughout the season. From these, farmers can quantify losses as well as market can manage prices. By adding a Digital Elevation Model (DEM) to this database, it is possible to estimate areas at risk of frost, providing subsidies to banks when contracting agricultural insurance.

5.8   Example of Data Integration: The Concept of Management Zones

In precision agriculture, one of the site-specific management approaches is the field division into smaller areas, which are called *management zones* (Khosla et al. 2008). This delimitation is carried out according to the crop yield potential of the different regions of the field and the factors limiting crop productivity, allowing the adoption of detailed and specific management practices for each zone, optimizing the return on investment.

Much has been said about this technique; nonetheless, it is important to note that the concept of management zones is different from the idea of homogeneous zones. In the latest, regions that are uniform in terms of one or a few parameters to be managed are delimited, for example, homogeneous zones for the application of fertilizers. However, the concept of management zones is broader than that, as the delimited regions must feature a similar combination of crop yield limiting factors, so that the entire management standard within the zones can be changed.

Thus, this approach allows many different site-specific managements, such as varying the cultivar planted in each region, changing the sown plant population and the soil fertility levels based on zone yield potential, and adapting pesticides and soil management based on the restrictions of each zone, among several other possibilities. Moreover, it is important to keep in mind that the information layers used in the analysis to define the management zones must be selected with sufficient technical criteria and must take into account a proper temporal window with different weather condition across the seasons.

Management zones are not immutable, because the factors that cause variability and limit productivity may change; however, there should be no frequent changes, or else, the benefits of the gradual suitability of management for each zone will never reach their full potential. Thus, the information layers that must be taken into account to delimit the management zones must show stability over time, that is, they must not be easily changed by management practices. Therefore, the chemical properties of the soil, such as the available phosphorus and potassium contents, are examples of what should not be used, since they can be corrected by variable rate of fertilizer application on the same growing season, which will cause the management zones to become obsolete.

The analysis and delimitation of the management zones can be carried out using several approaches, such as by simply overlaying maps with parchment paper and visually delimiting the regions, or based on data mining and artificial intelligence algorithms. However, the approach that is most commonly used is cluster analysis, which aims to create regions with low variability within the zone and, at the same time, high variability in relation to neighboring zones.

Several types of data have been suggested for consideration when adopting this strategy. In relation to the soil quality, properties such as texture, mineralogy and organic matter at different depths can provide insights on water holding capacity and natural fertility levels; likewise, important complementary information can be obtained. Proximal soil sensors are a common alternative, as they provide larger quantities of data in a cheaper way compared to traditional soil sampling, such as soil apparent electrical conductivity and magnetic susceptibility, which can contribute to the creation of more accurate maps.

In addition to soil proprieties, it is important to integrate data related to the crop, as it is the plants that express the response to the soil and other factors limiting production, such as soil fertility. Thus, it is important to integrate yield maps, obtained by sensors embedded in harvesters, in this analysis. A good yield mapping time series can bring a lot of information to variability interpretation and should be one of the main information layers.

An increasingly popular alternative in case this type of data is not available is to obtain remote sensing imagery. As the images obtained by satellites and drone-borne (unmanned aerial vehicles) have evolved a lot in recent years, allowing the estimation of vegetation indices, which generally express the plants' vigor and, consequently, can be used to understand how the crop is being influenced by other crop production factors.

When using productivity maps or images obtained by satellites or drones to define management zones, maps of several growing seasons should be considered, as crop yield may vary depending on several factors, which may not be recurrent in the same place from season to season. Using multiple crop seasons will eliminate noise and help clustering algorithms find recurring variability patterns in the area, thus being in line with the objective of the definition of management zones: finding areas with the same pattern.

There are numerous applications for data integration in GIS to obtain products used in decision-making processes, as mentioned earlier. Therefore, the analysts must consider the source of the data they intend to add to their database and then adequately analyze such data to obtain a reliable final product.

Abbreviations/Definitions

- Geographic Information Systems (GIS): It is a set of functions with advanced capabilities for storing, accessing, manipulating, and visualizing georeferenced information.
- GeoTIFF: It is a public domain metadata standard that allows georeferencing information to be embedded within a TIFF file.

- Brazilian Institute of Geography and Statistics (IBGE): It is the agency responsible for the official collection of statistical, geographic, cartographic, geodetic, and environmental information in Brazil.
- National Institute for Space Research (INPE): It is a research unit of the Brazilian Ministry of Science, Technology, and Innovations, responsible for fostering scientific research and technological applications and qualifying personnel in the fields of space and atmospheric sciences, space engineering, and space technology.
- China–Brazil Earth Resources Satellite (CBERS): It is a technological cooperation program between Brazil and China that develops and operates Earth observation satellites.
- TOPODATA: It is the Brazilian Geomorphometric Database.
- Structured Query Language (SQL): It is a standardized programming language used to manage relational databases and perform various operations on the data in them.
- QGIS: It is a free and open-source cross-platform desktop geographic information system application that supports viewing, editing, and analysis of geospatial data.
- Dark Object Subtraction (DOS): It is an image-based technique to cancel out the haze component caused by additive scattering from remote sensing data.
- Top of Atmosphere Reflectance (TOA): It is a unitless measurement that provides the ratio of radiation reflected to the incident solar radiation on a given surface.
- Bottom of Atmosphere reflectance (BOA): Also known as the surface reflectance, that is, satellite-derived Top of Atmosphere (TOA) reflectance corrected for the scattering and absorbing effects of atmospheric gasses and aerosols.
- Operational Land Imager (OLI): It is a remote sensing instrument aboard Landsat 8.
- Red, Green, and Blue space (RGB): It is any additive color space based on the RGB color model.
- Vegetation Index (VI): It is a spectral transformation of two or more bands designed to enhance the contribution of vegetation properties.
- Leaf Area Index (LAI): It is a dimensionless quantity that characterizes plant canopies. It is defined as the one-sided green leaf area per unit ground surface area in broadleaf canopies.
- Precision Agriculture: Precision agriculture is a management strategy that takes account of temporal and spatial variability to improve sustainability of agricultural production (International Society of Precision Agriculture—ISPA definition).
- Coordinate Reference System (CRS): It is a framework used to precisely measure locations on the surface of the Earth as coordinates.

Take Home Message/Key Points

- GIS is a powerful tool for integrating data for decision-making on crop management.
- A quality database is essential to generate reliable results.
- Data processing is a key step to extract the maximum of the database.
- Data integration provides knowledge to understand the phenomenon occurring in an area.

References

Burrough PA (1986) Principles of geographical information systems for land resources assessment. Oxford University Press, Oxford

Chaves JPS (1988) An improved dark-object subtraction technique for atmospheric scattering correction of multispectral data. Remote Sens Environ 24:459–479

Cróstra AP (2002) Processamento digital de imagens de sensoriamento remoto, 4th rev edn. Instituto de Geografia – UNICAMP, Campinas, 164p

Driemeier C, Ling LY, Sanches GM, Pontes AO, Magalhães PSG, Ferreira JE (2016) A computational environment to support research in sugarcane agriculture. Comput Electron Agric 130:13–19

Jensen JR (2007) Remote sensing of the environment: an earth resource perspective, 2nd edn. Pearson Prentice Hall, Upper Saddle River

Khosla R, Inman D, Westfall DG, Reich RM, Frasier M, Mzuku M, Koch B, Hornung A (2008) A synthesis of multi-disciplinary research in precision agriculture: site-specific management zones in the semi-arid western Great Plains of the USA. Precis Agric 9:85–100

Mather PM, Koch M (2011) Computer processing of remotely-sensed images: an introduction, 4th edn. Wiley Online Library, Oxford

Moreira MA (2011) Fundamentos do sensoriamento remoto e metodologias de aplicação, 4th edn. Editora UFV, Viçosa, 422p

Zanotta DC, Ferreira MP, Zortea M (2019) Processamento de imagens de satélite. Oficina de Textos, São Paulo, 320p

Chapter 6
Sampling and Interpretation of Maps

Tiago Cappello Garzella, Verônica Satomi Kazama, and Mario Hideo Sassaki

6.1 Introduction

As old as humanity itself is the practice of sampling, either by the spoon of the soup that the cook tastes to check its flavor or by the sip that the party's host tastes before offering the wine when the goal of the action is the same: knowing the whole by means of a small part. It was from the seventeenth century (Graunt 1662) that sampling began to be grounded in the way we know it today until it found most of its bases in the late nineteenth century with the emergence of the first methods of Modern Statistics. In fact, even nowadays, in most areas of science, the goal is to know in the best way possible a population, generally too large, from the investigation of some of its individuals.

Much of the decision-making on the management of a field is supported by sampling practices, whether considering soil analyses, observations taken during a field inspection, or considering investigations on the occurrence of a given disease. The larger the area, the greater the difficulty of making the correct decision. It is within this context that the set of tools presented by digital agriculture can favor the farmer, from the choice of the best sampling strategy to the interpretation of the results, making it simpler and faster to decide for the correct intervention in the field.

In this chapter, we present the basic concepts of sampling and the interpretation of results focused on digital agriculture, hoping that in the end the reader can choose the most appropriate strategy to study and know the production factor of his or her interest.

T. C. Garzella (✉) · M. H. Sassaki
APagri Consultoria Agronômica, Piracicaba, Brazil
e-mail: tiago@apagri.com.br; mario@apagri.com.br

V. S. Kazama
Federal University of Paraná, Curitiba, Brazil

D. Marçal de Queiroz et al. (eds.), *Digital Agriculture*,
https://doi.org/10.1007/978-3-031-14533-9_6

6.2   A Brief History

One of the earliest records on the use of sampling for scientific purposes dates back to 1662, when the British John Graunt described a method for estimating the population of London from partial information. In 1783, a French astronomer and mathematician Pierre-Simon Laplace estimated the population of his country using some principles of Thomas Bayes, a contemporary English statistician, who formulated what would soon be known as Bayes' theorem, which helped to substantiate the Modern Statistics. Although it seems very distant and even out of context, Bayesian inference, whose roots come from this time, served as a seed for computer intelligence or Bayesian learning, now widely used by digital agriculture when it comes to the use of machine learning techniques and artificial intelligence, including for sampling.

In agriculture, since 1920, there have been reports of collections of soil portions to test them for the concentration of nutrients that are important for plant growth. However, it was in the mid-1950s, when Modern Statistics had already been established, that this practice gained relevance, remaining virtually unchanged until the 1990s.

The predominant strategy until then consisted of collecting between 10 and 20 soil samples per uniform plot, with an area around 50 ha, then mixing them in a bucket to take a 300 g portion, called a composite sample, to be sent for laboratory analysis. It is evident that the most arduous task of this process was to assume the uniformity of the plot, ignoring that often there would be a mistake already at the beginning of the work, because the variability, even if at different intensities, is inherent to any field.

In the 1990s, with the emergence of precision agriculture, practical relevance is gained that allow variability to be considered in large-scale management of fields. Sampling, as an elementary approach to study the main factors of production, then begins to play a prominent role.

6.3   What You Cannot Miss About Sampling

Technically speaking, sampling is the process of representing a population (even more technically, a statistical population) by studying a portion (sample) of its individuals. If, for example, we want to know the percentage of plants in a soybean field that are being injured by a given pest, instead of evaluating each of them individually, it is much more appropriate to select some plants that represent as faithfully as possible the entire field. The way we select them is known as sampling, based on which we perform statistical generalization or inference. In order to obtain the percentage of attacked soybean plants, we take some samples from the area, perform an evaluation and extrapolate the result for the entire field, accepting some level of error.

Any deviation found between the actual result and that obtained by sampling, in Classical Statistics, is attributed to what we know as residual, uncontrolled or random factors that confuse the generalization process. As the soil, plants, and the vast majority of production factors are uneven and vary in space and time, it is important to use additional statistical procedures that consider and reflect these variations.

6.3.1 Sampling in the Context of Precision Agriculture

With the advance of mechanization and other facilities of modern technology, agriculture began to overuse generalization. Increasingly larger fields were managed based on a limited number of samples, assuming a uniformity of production that has never been observed in the field. This extensive practice continued for some period, but it became increasingly expensive, less efficient in terms of yield, and more harmful to the environment.

Motivated by these factors, researchers have made a growing effort in developing and applying field management strategies to make the agricultural activity more sustainable. The adoption of localized treatments of the crop, considering it as non-uniform, pointed to the most viable option for the maintenance of yield levels with reduction in the use of inputs, reaching much of what is conceptualized today as precision agriculture.

The most arduous task in this new context is how to identify and explain the behavior of the production factors that cause the nonuniformity often observed in the field. This identification and, consequently, the diagnosis and decision-making can only be accomplished based on information about the system, and the most appropriate way to obtain information, as seen earlier, is through sampling.

When this scenario of variability begins to be considered, it becomes evident that traditional statistical analyses, attributing to the residual the deviation found between the actual result and that obtained by sampling, no longer meet the purpose of precision agriculture. The alternative found to better represent the nonuniformity is the theory of regionalized variables, contained in the field of Spatial Statistics and established from 1950, in publications of Daniel Krige, studying gold mines in South Africa. The great differential of this type of analysis is to consider correlations between observations of neighboring samples, accepting the possible existence of a spatial component acting on the individuals studied.

Still, in our example of soybean plants, it is easy to accept that the behavior observed in samples collected at sites closer to each other tends to be more similar than that observed in samples taken at more distant sites. The key to success here is to determine the intensity of this principle and the distance up to which it acts for the characteristic under study, and the concept introduced in 1965 by a French engineer, Georges Matheron, known as Geostatistics, is used for this purpose. This is the study of the behavior of random variables that have a spatial structure.

Therefore, Geostatistics comprises the set of tools to determine the existence of spatial relationships between observations that make up a sample, making it possible to model spatial patterns, through a graph known as variogram; estimate values in unsampled sites, by means of generalization technique, or interpolation, known as kriging; obtain the uncertainty associated with an estimated value; and optimize the distribution of sampling points.

6.4 Sampling Strategies

There are basically two fundamental strategies for sampling used in precision agriculture: grid sampling and management zone sampling, from which some variations originate. In the first strategy, known as grid sampling or mesh sampling, georeferenced sampling points are systematically allocated throughout the area. The second strategy is carried out by management zones, also known as differentiated management units. In this strategy, prior to sampling, management zones must be defined based on production factors, such as yield, vegetation, or soil electrical conductivity maps, for example. The strategy, therefore, consists of creating clusters of regions in which historically the behavior of the crop is similar.

6.4.1 Grid Sampling by Point

This sampling is the most widely used strategy in Brazil to characterize the variability of the crop fields. By this method, a regular virtual grid is created over the study area, dividing it into polygons. A sampling point is generated within each polygon, and it can be located in the center or, randomly, inside it (Fig. 6.1).

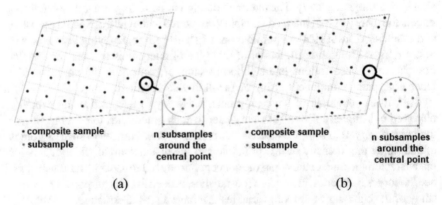

Fig. 6.1 Grid sampling with point in the center (**a**) and with random point in the cell (**b**). (Source: The authors)

The coordinates of the collection points are transferred to a Global Navigation Satellite System (GNSS) device, which will guide the collection team to each sampling site. Subsamples are collected within a radius around the point, of size to be defined by the user according to the purpose for which sampling is being carried out.

The larger the number of subsamples, the lower the sampling error, but the longer the time required for the work, and, consequently, the higher the cost. The subsamples are then homogenized, generating a composite sample, whose objective is to represent the point well.

The samples are sent for laboratory analysis, and then the results are linked to the respective coordinates. To generate the final map, the data are interpolated, a procedure that estimates values at non-sampled sites, hence filling the entire surface of the map.

An important factor for the planning of grid sampling by point, and perhaps one of the most frequent questions, is the grid size or sampling resolution that implies in the number of sampling points per unit area. The quantity directly affects the quality of the final map generated from the sampling. In general, the greater the number of points, the smaller the distance between them and the better the representation of the area. However, the costs with collection and analysis become higher.

What limits the adoption of low-resolution grid sampling is the distance between points. As presented in the chapter on spatial and temporal variability, to study the spatial pattern of a variable, the collection sites need to be within the limit distance at which the relationship between the samples still exists. If they are beyond this limit, the relationship will be random, making it impossible to properly model the spatial behavior and making the inference about unknown values much more subject to randomness.

Therefore, the aim is to obtain the lowest sampling resolution that does not affect the quality of the final map. And the most consolidated way to study spatial correlation and determine the limit distance from which spatial dependence ceases to exist is the use of the semivariogram (Yamamoto and Landim 2015).

By looking at the graph in Fig. 6.2, it can be noted that the semivariance tends to increase as the distance also increases until it stabilizes. From this point, known as range or amplitude, the semivariance changes little as a function of distance.

Since each parameter to be studied by sampling may exhibit different spatial behavior, it is necessary to construct a semivariogram for each parameter. The recommendation is that the grid size to be assumed should be the one which respects the half of distance limit (*range*), found by the semivariogram of the study variables. Better explaining, the correct is to perform a previous study of the area before establishing the best sampling intensity. In practice, what is observed is the adoption of intensities ranging from 0 to 5 hectare per point, choosing the inference method after evaluation of the semivariograms. If the observed range is smaller than the distance between the points or the model obtained has a low coefficient of determination, the data are not interpolated, and the map is constructed based on the mean of the values.

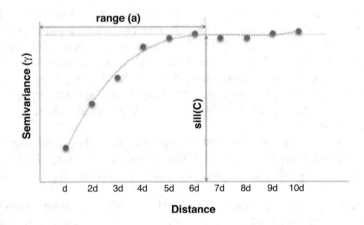

Fig. 6.2 Semivariogram with the components range (*a*) and sill (*C*). (Source: The authors)

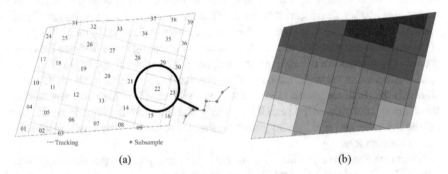

Fig. 6.3 Cell sampling with virtual grid on the plot and strategy for collecting subsamples in cells (**a**), besides the map obtained after processing (**b**). (Source: The authors)

6.4.2 Grid Sampling by Cell

Grid sampling by cell consists of dividing the field into subareas, not necessarily regular, called cells. Unlike point sampling, in this case, subsamples are collected along the entire area of each cell, seeking to represent not only one point, but the entire cell, as illustrated in Fig. 6.3.

The number of subsamples is recommended to be greater than that used for point sampling, since the area to be covered is much larger. There is no practical recommendation for the number of subsamples according to the cell area. It is important to also consider what is being sampled. For example, for soil sampling, from 10 to 20 subsamples are usually collected in each cell. It is evident that very large cells will require more than 20 subsamples. Usually, a zigzag walking is carried out in the collection of the subsamples. After homogenization, a composite sample is generated, and the result of its analysis is assigned to the entire cell area (Fig. 6.3b). For collecting field data with the SPAD chlorophyll index sensor for N dosage calculation for corn crop, it generally is recommended to take 30 measurements to represent each point.

To generate the map of the variable, interpolation is neither necessary nor recommended because the aim with this strategy is not to determine the spatial model. For plotting and final visualization, one must simply generate a color gradient according to the values obtained in each cell.

It can be noted that for this methodology, there is no concern about sampling size, since interpolation is not applied. Indeed, cell sampling is an alternative for those who intend to perform grid sampling but consider the required number of samples in the point method to be very high. The major limitation of this strategy is that it does not consider spatial dependence, being just a conventional sampling with greater detail.

6.4.3   Targeted Grid Sampling

In targeted sampling, there is no regular distribution of sampling sites. These are chosen based on maps already obtained from the area, seeking to investigate specific sites in the plot.

The main objective of this strategy is to increase sampling resolution in regions where greater variability is observed and reduce it in more uniform regions. In this way, it is possible to have higher final grid size of the area, without compromising the analysis of spatial behavior and the quality of the interpolation and, consequently, of the map to be obtained. Figure 6.4 shows a representation of this sampling strategy.

To define sampling points, it is recommended to use maps that represent a history of the behavior of the crop or variable under study in the field. In general, yield

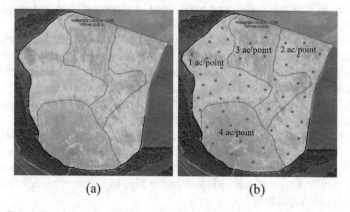

Fig. 6.4 Targeted grid sampling strategy, with the delimitation of zones with different spatial behaviors (**a**) and allocation of different sampling resolutions according to the variability observed in each zone (**b**). (Source: Technical material from *APagri Consultoria Agronômica*)

maps and biomass variability maps are used; however, maps of soil physical attributes, such as relief maps and maps of soil type, texture, or electrical conductivity, can also be used.

It is important to highlight that not necessarily areas with higher production potential will receive fewer points, while areas with lower production potential will receive more points. What usually defines the direction of points and supports the scientific basis of the method is variability, that is, regions with higher variability require greater sampling resolution, to represent the spatial behavior.

Once the points have been established, the methodology for collecting the samples is identical to that of grid sampling by point.

6.4.4 Sampling by Management Zones

A more recent trend that is likely to be the path of the future, especially for more mature production systems, is sampling by management zones, or site-specific management units (SSMUs).

Two approaches are adopted using this sampling strategy. The first one aims, from an observed pattern of variability, to virtually delimit the regions and investigate each of them in search of the causative agent of the difference.

The second, more common, seeks to group regions of the same plot that have similar behavior, that are stable over time, and that characterize their production potential. Therefore, in addition to the variability in space, the temporal stability of the observed patches is an important characteristic of the strategy, since the observed behavior is expected to represent permanent characteristics of the area, not caused by management actions.

Typically, management zones are obtained by the composition of several maps, involving from soil physics data such as relief, depth, and electrical conductivity, to the history of satellite images and yield maps. Any information that is useful for grouping similar regions of the field can be considered.

Clustering techniques are used for creating map composition (yield map, soil electrical conductivity, vegetation index, etc.) to design the management zones. A widely employed clustering technique is to use the k-means algorithm. These techniques are addressed in the chapter on geographic information system (GIS).

After identifying the zones, each of them can receive individualized treatment, from sampling, choice of cultivar, management, harvesting, to accounting. The sampling procedure is identical to that of grid sampling by cell, as illustrated in Fig. 6.5. Thus, subsamples are collected in such a way to represent the entire unit, homogenized, and their composite sample is taken to the laboratory. The result is then assigned to the entire unit, without the need for any process of statistical inference, such as interpolation.

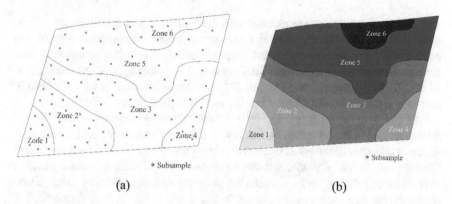

Fig. 6.5 Graphical representation of sampling by zones, with the subsample points (**a**) and the final map, in which the result of each analysis is associated with the corresponding zone (**b**). (Source: The authors)

6.5 Map Generation

The production factors of a field can be represented by thematic maps, in order to assist in the interpretation and decision-making throughout the agricultural cultivation. The principle behind the obtaining of maps is generalization, which, as already discussed, aims to infer about values at unknown sites, using values at known sites.

The main methods for constructing maps are interpolation or generalization from an average value of a zone or region. The selection of the method takes into account the sampling strategy chosen, according to a previous study.

6.5.1 By Interpolation

From sampling points, it is possible to reproduce the characteristics of a spatial variable for unsampled regions by interpolation. The estimation for a given unsampled position is performed by mathematical functions and can take into account only the nearest sampled points or the entire set of samples. The interpolation accuracy is influenced by the number of sampling points and how they are distributed in the area (Yamamoto and Landim 2015).

According to Li and Heap (2014), the methods for spatial interpolation of biophysical attributes for Environmental Sciences can be classified into three categories: (1) non-geostatistical methods, (2) geostatistical methods, and (3) combined methods. The first category uses deterministic models that are based on Euclidean distances between the sampling points and does not consider the uncertainty of the estimation of unsampled points. The main example is the inverse distance weighting (IDW) method. On the other hand, geostatistical methods take into account the spatial structure of the sampled values, using stochastic models that consider the uncertainty of the estimate associated with the estimator. The main example of this method is kriging. Finally, the third category is based on non-geostatistical and geostatistical approaches or other statistical methods for estimating unknown points, such as the method of regression trees combined with kriging (Li and Heap 2014; Li et al. 2019; Yamamoto and Landim 2015). The most commonly used methods in precision agriculture are ordinary kriging and IDW (Betzek et al. 2019).

The choice of the most appropriate interpolation method depends on a number of factors related to the sampling and nature of the data (Santi et al. 2016). However, the most common approach is to choose kriging when there is an adequate characterization of the spatial model and inverse distance weighted when it is not possible to model the spatial behavior or when the sampling distances are very small, eliminating the need for modeling. For example, when grid sampling is performed but, at the end of the process, the model obtained for a given attribute does not have a good fitting coefficient, interpolation by IDW is usually applied. It is important to emphasize that, by choosing this method, one assumes a uniform spatial pattern in the area, hence most likely making an error, which may or may not be accepted according to the degree of risk aversion by the person responsible for the analysis. The best solution in a situation like this, which minimizes the error, is to opt for using the average value of the attribute in the area.

Another very common situation is when there is a high intensity of information about a variable, as in the cases of the use of harvesters with yield mapping systems or the use of soil electrical conductivity sensors, active canopy sensors, or even aerial or orbital imaging. In these cases, the values are obtained every few meters, and the interpolation strategy aims only to fill the empty spaces in order to create a surface. In such a situation, IDW is more recommended than kriging.

6.5.2 By Management Zones

Especially in the spatial mapping of soil attributes, there is often the presence of regions that are grouped by homogeneous characteristics and have little change over time. As an alternative to the management of these areas, the field can be mapped by zones, as shown in Sect. 6.4.4.

After designing of the zones and sampling, the results are associated with the entire zone. In this process, generalization does not involve inference and, at most, can be performed by the average, when the same zone has more than one result.

The main advantage of this mapping strategy is to allow a reduction in the number of samples for generating the maps, minimizing the costs with collection and laboratory analysis. As modeling and interpolation are not necessary, in general the process of obtaining the maps is also simpler and faster.

6.5.3 Crop Variability Mapping

A fundamental component of the concept of precision agriculture is the ability to measure spatial variations in soil factors and evaluate their influence on crop variability, as well as understand the variation between cropping seasons to apply appropriate management strategies (Earl et al. 2003). In this context, it is necessary to evaluate, quantify, and map the variability of the field, seeking to increase the efficiency in the use of available resources. Considering spatial and temporal variability is, therefore, essential to achieve the best management practices, involving the application, at each specific site, of the agricultural input at the correct dose and at the right time (Gebbers and Adamchuk 2010).

The study of field variability is considered the first step to perform precision agriculture (PA) because, if there is no significant spatial and temporal variability in a crop, there may be no reason to apply PA. The main source of information to visualize the spatial change of crops is the yield map, capable of showing the response of the crop to the management adopted between seasons and to the variation of the production factors (Molin et al. 2015; Earl et al. 2003). Yield can be obtained by direct and indirect methods. In the direct method, data are obtained through sensors coupled to the harvesters, which can measure the flows of grains and other solid agricultural products. For a more complete analysis of field variability, yield maps can be combined with other layers of information, such as elevation maps and soil chemical and physical properties.

In certain applications, the mapping of yield through the direct method can be considered a late information, for example, in cases where production needs to be monitored during the crop cycle. Thus, indirect methods become an interesting alternative due to the speed and ease of measuring large tracts of areas in various stages of the crop. In this type of sampling, the information can be estimated from images collected by orbital, aerial, or terrestrial platforms. These data make it possible to generate useful maps, such as that of normalized difference vegetation indexes (NDVI), providing information on differences in field biomass that can express the vigor or yield of certain crops (Xue and Su 2017). The interpretation of NDVI values can also indicate crop diseases, water stress, presence of pests, nutritional deficiencies, and other relevant conditions that affect the yields of the fields (Pôças et al. 2020).

The common procedure from vegetation maps is to look for which factors are causing crop patches. Therefore, some sites are allocated for field inspection, and once the causative agent is understood, an indirect map of this factor can be created, considering the variability verified by the vegetation map. This strategy can serve to delimit problems of nutritional deficiency, phytosanitary problems, presence of weeds, poor drainage, soil compaction, equipment, or operator failures, among others. The importance of mapping these factors is to allow for localized action, either for control or for adjusting management practices to the specific production potential of each region.

6.5.4 Mapping of Soil Attributes

Without knowing soil variability, it is virtually impossible to study field variability. Therefore, generating maps of soil attributes is fundamental for agronomic management. In general, the adoption of precision farming practices in a property begins precisely by the mapping of soil attributes, aiming to ensure a good basis for crop development. Soil maps are important for agricultural management because they enable the optimization of field operations, such as tillage, correction, and fertilization, and assist in environmental monitoring, land use planning, and conservation of natural resources. Among the PA practices performed in soil management, fertilization can be considered the main one.

To recommend these practices, the most common form of mapping is still by grid sampling. As already discussed in this chapter, to produce reliable maps, one cannot neglect the grid size, that is, the collection points cannot be arranged beyond the limit distance at which there is a relationship between them. In opting for low sampling resolution, it is not likely that models for estimating unknown values will be able to return good results.

Very short limit distances are often verified, especially for the attributes phosphorus and potassium, requiring high sampling resolution and the consequent high cost with soil collection and analysis, which can make grid sampling unfeasible. Therefore, there is a current trend of keeping the strategy in grid only as exploratory sampling of an area, and that the targeted strategies or by management zones are preferred for properties with a more established history.

One way to minimize the number of samples needed is to use maps with previous information on soil spatial variability, which makes it possible to identify more representative sampling sites in the field. Thus, maps of soil apparent electrical conductivity have great potential for assisting in sampling planning, because they indicate the intrinsic variability of the soil and have high spatial resolution.

Another variable that is highlighted is the topography, which also helps explain soil variability, assisting in sampling planning. Mapping the relief is easier and costs much less compared to the mapping of soil attributes (Sanchez et al. 2012). It is important to highlight that the apparent electrical conductivity of the soil also responds to this trend of relief.

In the stages of both sampling planning and inference, the use of satellite imagery or soil electrical conductivity can help, especially in the mapping of attributes such as clay and organic matter and others that are related to these variables. One approach that can gain relevance with the adoption of more robust processing libraries is to consider this type of information as a covariate when estimating soil attributes, increasing the amount of information available about each of them and the final quality of the map.

Digital soil mapping methodologies such as these can be associated with the conventional approach, improving the mapping of soil attributes, at both small and large scales (Sayão et al. 2017).

6.5.5 Mapping of Pests and Diseases

It is still common to try to explain the variability observed in the field only by investigating soil fertility. However, most often, the causative agent of the differences is associated with other factors of production, especially those that are related to the plant's ability to absorb water. Phytosanitary problems, many of which do not always cause obvious symptoms, can go unnoticed and even cause major damage and yield losses. Therefore, as important as understanding the bases of the soil on which the crop is planted is to study the occurrence and impact of pests and diseases.

The main objective of identifying the phytosanitary problem and its severity and extent is to propose a localized treatment that increases the efficiency of control economically and environmentally. Although this type of mapping requires sampling techniques already presented in this chapter, the high variability that certain pests and diseases have in the field ultimately requires higher sampling resolution. For choosing the best strategy, it is essential to know the behavior of the causative agent, its mobility in the field, its population fluctuation, and its level of damage because, although many of them cause symptoms in patches, it is common to note impacts on yield even when there are no symptoms in the plants.

Examples of problems that can remain hidden for several cycles are nematodes, which parasitize plant roots and cause symptoms that are easily confused with nutritional problems or with other pests. This characteristic makes diagnosis difficult and results in major damage to the crop.

The most used strategies for pest mapping are grid sampling and targeted sampling (Fig. 6.6), using biomass variability maps as the basis for establishing the collection sites. The most relevant point approach can also contribute to monitoring, and once the occurrence is detected, a more intense evaluation process begins.

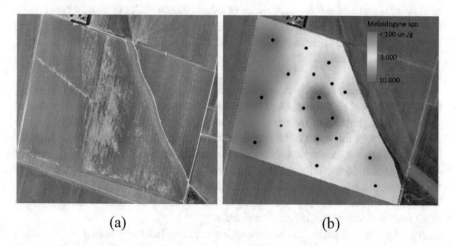

(a)   (b)

Fig. 6.6 Biomass mapping using unmanned aerial vehicle (**a**) with targeting of sampling points and the final map obtained (**b**) for the diagnosis of nematode occurrence. (Source: Technical material from *APagri Consultoria Agronômica*)

Sampling may involve laboratory analysis of both soil and plant tissue—depending on the pathogen—and visual evaluation, assigning a score related to the damage or level of infestation verified. Combined methods are also adopted, by sending some samples to the laboratory and using more intense grids for visual evaluations.

An interesting solution is to interpolate the results of laboratory analyses, which have lower sampling resolution, considering the results of visual evaluations as covariates, which are sampled in greater intensity. Thus, the laboratory cost is minimized without compromising the quality of the spatial modeling.

Some care should be taken to facilitate the standardization of visual evaluation, with the addition of references or photos, which used as a comparison facilitate the attribution of a score to the sample. There is also a tendency to assess image recognition, using neural networks and artificial intelligence. Thus, from a sample photo, the algorithm itself returns a score.

Zone sampling can also be used, identifying, mainly by satellite or drone images, patches in the development of the field. From the virtual contour of the patches, or zones, samples representing each of the zones are collected, and the result will be the observed mean.

6.5.6 Interpretation of Maps for Decision-Making

The objective of any mapping is to enable the study of the spatial behavior of a given variable, to make decision-making more assertive in the management of fields. However, representing the main factors of production through digital maps alone is not always sufficient to establish a cause-effect relationship with crop yield.

The step of analysis and interpretation sometimes comes up against the large amount of data to be considered, many of which interact with each other and make it very arduous to identify the attribute most responsible for yield. When strong relationship of cause and effect is not found, it is common for the user to lose motivation and the perception of the value of technology, returning to conventional agricultural practices.

Any investigative activity needs three fundamental bases. The first is inference, through statistical or geostatistical analyses, which seeks to know, understand, and validate the arguments of the context in which the problem is inserted. The second is opportunity, in the sense of having conditions or resources to carry out the investigative action. For example, some variables can be difficult to study, due to either extremely variable behavior or very high cost of analysis. The third base, probably the least explored, is intuition. Interpretation usually requires something more than looking at numbers and maps and needs to overcome the empirical, allowing one to think and read more than is being presented.

All three bases mentioned have their importance, but the last two are the most difficult to perform, because they do not allow the adoption of a process through which the desired result is obtained, once it is executed. Therefore, in this chapter, the focus is on the inference step.

In the interpretation of the maps, the first step usually consists of exploratory analysis of the data, through descriptive statistics. The objective is to describe and summarize the data set, verifying the levels of adequacy of the main attributes to the requirements of the crop. And each map can receive a legend pattern that makes it easier to visualize regions with critical levels.

In this step, the analysis may involve the results considering the interpolation process or be restricted to the points at which there is data for most attributes, such as a soil sample sites. In this case, variables that are in another sampling pattern can be rearranged (or, in the most common term for the procedure, resampled), so that the value is obtained at the same site of the soil points.

As vegetation maps or yield maps usually have more points with information than soil sampling, the procedure adopted here is to obtain the average of the values contained within a given distance around the point with result of soil. The goal is to group the information into the same spatial pattern.

After exploratory analysis, it is recommended to perform correlation analyses, which assess the degree of association between the variables. By organizing the correlation coefficients (r) in a table, one arrives at what is known as a correlation matrix, whose tabular form enables the simultaneous analysis of the associations, as well as the identification of the ones that are most relevant and that can already explain part of the result (Fig. 6.7).

Another analysis that has received increasing attention in the context of digital agriculture is multivariate analysis. As the farmer increasingly faces a large number of different variables in the analyses—whether considering different attributes or

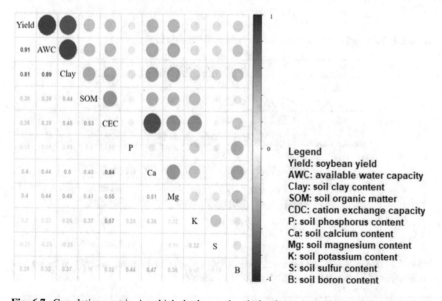

Fig. 6.7 Correlation matrix, in which the larger the circle, the greater the correlation, with blue indicating positive correlation and red indicating negative correlation. (Source: Data obtained from the *APagri* information base)

considering several years of information history—multivariate data analysis techniques that facilitate interpretation have great value. Therefore, multivariate analysis consists of a series of procedures to simplify the study of the association of two or more sets of measurements performed in the samples.

Among the multivariate analysis techniques, principal component analysis (PCA) can help the farmer by enabling him or her to explain as much information as possible with a small number of principal components (PC). The technique allows replacing the original variables, grouping them into new uncorrelated variables, which contain most of the information of the original ones. By this method, it is possible to reduce the number of variables evaluated, highlighting those that are more important to explain the variability of the area. Each new variable, called principal component, is formed by a function that considers a set of original variables (Table 6.1).

Typically, with the first four principal components, it is already possible to explain more than 90% of the variability of an area, as illustrated in Table 6.2.

As examples of application, Santi et al. (2012) used this type of analysis to study the spatial and temporal variability in the yield of grain crops from soil chemical and physical attributes, and Oliveira et al. (2002) applied principal component analysis to define management zones.

The growing number of libraries and packages, many of which for free, especially for R and Python languages, facilitates the use and adoption of advanced analysis techniques to simplify the interpretation and decision-making of agronomic management.

Table 6.1 Principal components of the analysis with the data in Fig. 6.7

PC	attribute/coefficient			
1	S	P	K	B
	0.384551	0.17115	0.09572	-0.17625
2	CEC	K	SOM	Ca
	0.463242	0.450939	0.293998	0.284008
3	Yield	AWC	Clay	S
	0.290506	0.26248	0.182628	0.150755
4	K	P	SOM	Clay
	0.648207	0.519532	0.223757	0.185405

Table 6.2 Importance of the principal components (PCs) of multivariate analysis, showing that the first four PCs already account for 91% of the variance

Importance of principal components	PC1	PC2	PC3	PC4	PC5	PC6	PC7	PC8	PC9	PC10
Standard deviation	2.32	1.55	1.17	0.92	0.72	0.54	0.38	0.16	0.12	0.06
Proportion of variance (%)	49%	22%	12%	8%	5%	3%	1%	0%	0%	0%
Accumulated proportion (%)	49%	71%	83%	91%	96%	98%	100%	100%	100%	100%

Abbreviations/Definitions

- Inference: A process of drawing conclusions about an underlying population based on a sample or subset of data.
- Geoestatistics: A class of statistics used to analyze and predict the values associated with spatial or spatiotemporal phenomena.
- Semivariogram: A geostatistical function that describes the spatial correlation model of a set of data.
- Semivariance: A property that expresses the degree of relationship between points in an area.
- Range: A parameter of a semivariogram that indicates the distance at which the model first flattens out. From this distance there is no spatial correlation between the data.
- Sill: A parameter of a semivariogram that indicates the semivariance value at which the model first flattens out.

Take Home Message/Key Points

- The main challenge of the sampling is how to represent the whole as much as possible, using as few observations as possible.
- The objective of any mapping is to enable the study of the spatial behavior of a given variable, to make decision-making more assertive in the management of fields.
- The key in obtaining maps is in the generalization process, by which aims to infer about values at unknown sites using values at known sites.
- Any investigative activity needs three fundamental bases: inference, opportunity, and intuition.

References

Betzek NM, De Souza EG, Bazzi CL, Schenatto K, Gavioli A, Magalhães PSG (2019) Computational routines for the automatic selection of the best parameters used by interpolation methods to create thematic maps. Comput Electron Agric 157:49–62

Earl R, Taylor JC, Wood GA, Bradley I, James IT, Waine T, Welsh JP, Godwin RJ, Knight SM (2003) Soil factors and their influence on within-field crop variability. Part I: field observation of soil variation. Biosyst Eng 84(4):425–440

Gebbers R, Adamchuk VI (2010) Precision agriculture food security. Science 327(5967):828–831

Graunt J (1662) Natural and political observations mentioned in a following index, and made upon the Bills of Mortality. London Repub Introduction B. Benjamin J Inst Actuaries 90:1–61, 1964

Li J, Heap AD (2014) Spatial interpolation methods applied in the environmental sciences: a review. Environ Model Softw 53:173–189

Li J, Nearing MA, Nichols MH, Polyakov VO, Guertin DP, Cavanaugh ML (2019) The effects of DEM interpolation on quantifying soil surface roughness using terrestrial LiDAR. Soil Tillage Res 198:104520

Molin JP, Do Amaral LR, Colaço AF (2015) Agricultura de precisão. Oficina de Textos, São Paulo

Oliveira A, Franca G, Avelar G, Mantovani E (2002) Análise de componentes principais para definição de zonas de manejo em agricultura de precisão. In: Congresso Nacional De Milho

E Sorgo – Meio ambiente e a nova agenda para o agronegócio de milho e sorgo, 24., 2002, Florianópolis. Resumos expandidos. ABMS: Embrapa Milho e Sorgo/Epagri, Sete Lagoas/ Florianópolis

Pôças I, Calera A, Campos I, Cunha M (2020) Remote sensing for estimating and mapping single and basal crop coefficients: a review on spectral vegetation indices approaches. Agric Water Manag 233:106081

Sanchez RB, Marques Júnior J, Pereira GT, Baracat Neto J, Siqueira DS, Souza ZMD (2012) Mapeamento das formas do relevo para estimativa de custos de fertilização em cana-de-açúcar. Engenharia Agrícola 32(2):280–292

Santi AL, Amado TJC, Cherubin MR, Martin TN, Pires JL, Flora LPD, Basso CJ (2012) Análise de componentes principais de atributos químicos e físicos do solo limitantes à produtividade de grãos. Pesq Agropec Bras Brasília 47(9):1346–1357

Santi AL, Giotto E, Sebem E, Amado TJC (2016) Agricultura de precisão no Rio Grande do Sul. CESPOL, Santa Maria

Sayão VM, Demattê JA, Bedin LG, Nanni MR, Rizzo R (2017) Satellite land surface temperature and reflactance related with soil. Geoderma 325:125–140

Xue J, Su B (2017) Significant remote sensing vegetation indices: a review of developments and applications. J Sens 2017:1–17

Yamamoto JK, Landim PMB (2015) Geoestatística: conceitos e aplicações. Oficina de Textos, São Paulo

Chapter 7
Application of Drones in Agriculture

**Lucas Rios do Amaral, Rodrigo Greggio de Freitas,
Marcelo Rodrigues Barbosa Júnior, and Isabela Ordine Pires da Silva Simões**

7.1 Introduction

Drone history faces a contrast between the previous high technology development for military purposes since a century ago and the present when civilians use it for multiple purposes such as recreational activities and providing professional services. The drone market expanded worldwide, exploiting various businesses, and the agricultural drones aroused the interest of major companies as the Chinese DJI and XAG and the Japanese Yamaha, among others.

The popularization by increasingly adopting such equipment forced international air space control and regulatory authorities to establish standards and concepts to keep the security of previous air activities. That job also aimed to stick the aeronautical culture of airspace use to beginners. All vocabulary to nominate equipment, parts, accessories, and rules were locally established by countries considering the International Civil Aviation Organization (ICAO) orientations. The common sense over the varied nomination of such equipment is that no pilot is on board, a no crewed aircraft. Besides the popular term drone, others like unmanned aircraft (UA), unmanned aerial vehicle (UAV), and remote piloted aircraft (RPA) are employed in varied situations. The ICAO technically classifies unmanned

L. R. do Amaral (✉) · I. O. P. da Silva Simões · R. G. de Freitas
School of Agricultural Engineering, University of Campinas, Campinas, SP, Brazil
e-mail: lucas.amaral@feagri.unicamp.br; i117271@dac.unicamp.br; r228594@dac.unicamp.br

M. R. B. Júnior
Department of Engineering and Mathematical Sciences, School of Agricultural and Veterinary Sciences, São Paulo State University (Unesp), São Paulo, Brazil
e-mail: marcelo.junior@unesp.br

© The Author(s), under exclusive license to Springer Nature Switzerland AG 2022
D. Marçal de Queiroz et al. (eds.), *Digital Agriculture*,
https://doi.org/10.1007/978-3-031-14533-9_7

equipment into three categories: RPAs, model airplanes, and autonomous equipment. The RPAs and models have a ground station control that can interfere with the flight anytime, even when there is a computer system control. However, models must be applied only to recreational activities. While the autonomous equipment does not allow pilot intervention after flight starts. Such equipment is prohibited in some countries. We believe the appropriate term for representing such equipment for agricultural purposes is "RPA," and therefore, it will be our choice in this text.

RPA use requires a system composed of a ground control station, the communication link, a way to interact with the air space controller and other aircraft, and the unmanned aircraft itself. The system is the so-called RPA system (RPAS) or UAV system (UAVS). The ground control is a group of equipment necessary to fly the RPA, usually represented by the remote control and flight control software. The communication link establishes telemetry among RPA and control stations. The communication to the air space controller may be by radio or mobile phones. Not less important is the team to operate the system. It is recommendable that an auxiliary person observes the RPA, and it is obligatory under some special conditions and in some locations. The observer helps the pilot visualize the RPA against air traffic, barriers over the flight route, or anomalous RPA behaviors.

It would be possible to classify RPAs from many characteristics as engine type, workload, and purposes. However, a simple way is to differentiate into two main flight characteristics: RPA fixed wings and rotor. Fixed wings seem like small airplanes and generally flies under higher speeds to air support and achieve more extensive flight autonomy. Launching systems enable takeoffs by a simple throw, and parachutes ensure smooth landing without damaging such equipment. RPA rotorcraft are generally multirotor and present two to eight rotors allowing them to fly at low speed, improving maneuverability. Flight autonomy is fewer than fixed wings. However, they are suitable to operate in places with obstacles or small experimental plots, and takeoff and landing are easy. It is often used to spread agricultural supplies as the easiness of uploading agri-chemical solutions or bio-products. Hybrid equipment tries to embrace the best of two previous types as they vertical takeoff and land while flying as a typically fixed-wing, reaching significant autonomy. For all this, the best equipment choice should consider the operation goal. Flight autonomy is one of the main bottleneck since electric power engines are predominant. Presently, it is possible to find large RPAs, especially sprayers, using combustion engines or hybrids to avoid such disadvantages.

7.2 Legislation

Official organizations are in charge of each country to organize and legislate about RPAs. They are establishing rules to ensure security for RPAs and other aircraft. The new regulations are dealing with the following: (1) RPA characterization regarding load capacity and engineering project approval; (2) RPA and pilot habilitation and registers; (3) legal aspects, such as insurances; (4) the communication

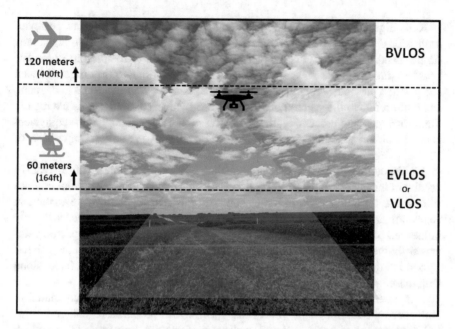

Fig. 7.1 Air traffic zones by operation altitude and respective visual categories. *BVLOS* beyond visual line of sight, *EVLOS* extended visual line of sight, *VLOS* visual line of sight

between RPA, air space controller, and, in some cases, to other aircraft; (5) flight plans according to regional specificities, such as altitude restrictions, risk zones, and visual contact to the RPA; and (6) type of operation and technical prerequisites.

Following the RPA flight is essential to a safely operation. The pilot and the auxiliary human observer should both be responsible for a visual contact to the RPA. In situations that visual contact to the RPA is possible without any auxiliary equipment, the flight is assigned as visual line of sight (VLOS; Fig. 7.1—Brazilian rules). When direct visual contact is not possible and a human observer or an auxiliary equipment must be present to keep safe, the flight is classified as extended visual line of sight (EVLOS). Another category is the beyond visual line of sight (BVLOS), to a situation that visual contact is not possible even with auxiliary assistance. The BVLOS flight is a particular category as it splits the air space with other aircraft and usually demands specific certifications from the RPA and pilot in any country.

7.3 Sensors

The main element to be considered in most agricultural monitoring is the sensor embedded in the RPA. In this sense, it is necessary to consider sensor spectral characteristics as a function of the desired survey/monitoring and their dimensions and weight, as RPAs have legal and design restrictions on transport.

RGB digital cameras are the most common sensors used, similar to personal/professional digital cameras or smartphone cameras. To be coupled in RPA, they simply need to be fitted with trigger devices connected to the aircraft controller board so that images are taken at the correct time according to the aircraft's speed and position previously established in the flight plan. The acronym RGB (red, green, and blue) refers to the spectral bands registered by the sensor in the visible region. Thus, during image processing, the respective colors are assigned to the equivalent bands to obtain true-color images (Fig. 7.2) similar to the color composition of human vision.

Multispectral cameras have at least one filter beyond the visible spectrum (RGB), usually in the near-infrared (NIR) region. The NIR band allows other color compositions like false-color and different compositions that highlight the variation in vegetation and calculate vegetation indices (IV) since the NIR reflectance is heavily influenced by the canopy of plants (Fig. 7.3). In addition, more multispectral cameras with five bands are becoming available, adding the red-edge region, which has a good relationship with vegetation vigor. Furthermore, some cameras have, along with these, a thermal infrared filter.

It is essential to mention that many users opt for RGB cameras compared to multispectral ones due to the lower cost. Thus, several proposed vegetation indices (VI) for this type of sensor emerged, such as the triangular green index (TGI), which is sensitive to leaf chlorophyll content. However, some users have modified RGB cameras, removing the blue band and including near-infrared, desiring to use more traditional VI (such as NDVI) due to their greater sensitivity to vegetation. This procedure has reservations, given two main reasons: (1) with the absence of the blue band, it is not possible to generate images in true-color composition, which can compromise photointerpretation, and (2) the reflectance measured by the modified infrared band may present variations in its calibration (gray level), which negatively impacts the reflectance calculation, compromising the vegetation indices

Fig. 7.2 RGB images obtained with RPA in sugarcane area (left) and orange orchard and pasture area (right)

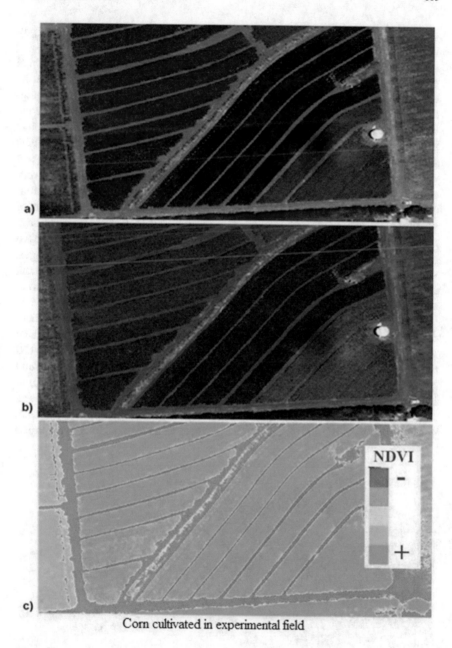

Corn cultivated in experimental field

Fig. 7.3 Image obtained by RPA with multispectral sensor: (**a**) Composition in true-color with RGB bands; (**b**) false-color composition with NIR, R, and B bands; (**c**) image with vegetation index (NDVI)

calculation, and consequently the reproducibility of the measurements and interpretation over the time (Nijland et al. 2014; Zhang et al. 2020).

Hyperspectral remote sensing or imaging spectroscopy, either through orbital platforms or by RPAs, is still little used due to its greater complexity and cost. Its use is mainly in the early stages of research and development. This type of equipment uses the same principles as reflectance spectroscopy, with each pixel in the image corresponding to hundreds of narrow-spectrum bands, composing an image in hypercubes (Fig. 7.4). This type of image usually has in the order of gigabyte sizes and, as multidimensional data, needs machine learning algorithms to detect patterns (Fig. 7.5). One of its primary uses is to study the spectral behavior of vegetation, aiming to develop combinations of specific wavelengths (vegetation indices). Another difficulty to use a hyperspectral sensor in RPAs is the spectral sensitivity, as they work with very narrow spectrum bands. It is necessary to have more excellent stability and precision in the positioning of the aircraft to record the reflectance data separately in hundreds of wavelengths without "blurring" images, which can be a limiting factor. Hyperspectral cameras enable the study of different spectral patterns of plants' physiological and structural properties when applied to vegetation monitoring.

Successful multi- and hyperspectral imaging requires some extra care and procedures, mainly to appropriately obtain the reflectance. Reflectance values result from incident light's recorded data (which hits the target—usually solar irradiation) and the energy reflected by the targets at different wavelengths. Therefore, multi- and hyperspectral sensors should have, in addition to the reflectance sensor (camera), an incident light sensor (solar sensor; Fig. 7.6a). The irradiance measured by the solar sensor is strongly dependent on its orientation to the sun during flight. For example, pointing directly at the sun will measure a different value than pointing to the top of

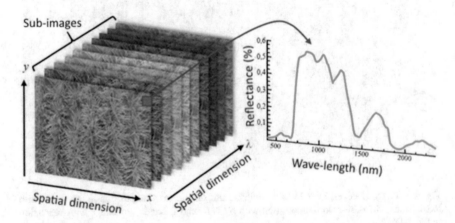

Fig. 7.4 Example of a hyperspectral image of green leaves. Stacking images in narrow bands forming a 3D hypercube. Spectral dimension (number of bands) (λ). Spatial dimensions (x, y). The graph exemplifies the reflectance of a pixel along the reflected electromagnetic spectrum recorded by the sensor

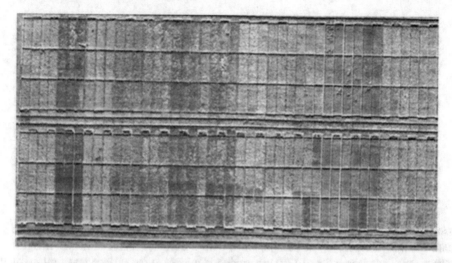

Fig. 7.5 Typical hyperspectral image of a barley field, showing a level of plant variability not observed with the naked eye. (Source: public data from the internet)

Fig. 7.6 Example of RPA with a coupled camera (downward facing), which will allow the calculation of the reflectance value of the imaged surface (**a**) and the reflectance panel (**b**)

the sky (when nadir). That is why the aircraft magnetometer must be correctly calibrated, providing accurate direction and orientation information, so that specialized processing software (such as Pix4Dmapper or Metashape) can provide the best lighting corrections according to the position of the aircraft and the sun. These incident light sensors are also effective in correcting for changes in the lighting conditions of the imaged area, such as when the sky is completely cloudy and the irradiance changes during flight. Following such warnings, collecting images on different days can be analyzed together with varying conditions of lighting.

However, this device cannot correct reflectance on days when the sky is partially cloudy with clouds shading part of the area but not the sensor. Another good practice is to fly around solar noon, avoiding shadows that affect the registered reflectance, whether from the plants themselves, surrounding trees, or any other obstacle.

In addition to the solar sensor, it is also recommendable to use a calibrated reflectance panel whenever data is collected (Fig. 7.6b). The panel usually supplied by the sensor manufacturer provides a reliable measure of irradiance on the ground, helping to confirm the information collected by the solar sensor. An image of the panel is acquired before the flight with the reflectance sensor placed on the ground. Thus during the processing of the collected images, it can perform compensations on the registered reflectance. However, its use is not recommended on days with partially cloudy skies precisely because it only represents the lighting in the place positioned, while the lighting will vary according to the clouds.

Another type of sensor that has been arousing great interest for agricultural use is thermal cameras. These sensors work with reflectance recorded at wavelengths generally between 7.5 and 13.0 µm, which is related to the surface temperature of the targets. Figure 7.7 shows two examples of thermal images used to distinguish targets with very different temperatures, such as animal monitoring (left) and the canopy variations (right) caused by plant stress, such as water deficiency.

A depth sensor called a LiDAR (Light Detection and Ranging) is a different type of camera. An active sensor system emits its signals to the earth's surface, using wavelengths ranging between 0.7 and 1000 µm (within the near-infrared region), and records the reflected signal, similar to radar technology. The time-lapse determines the distance between sensor and targets using the emitted laser pulse and the signal reflected independent of environment light. The LiDAR on RPAs allows the production of digital terrain models (MDT) and surface models (MDS) through point clouds generated with the data recorded by the sensor. Along with the aircraft position information, these points will register coordinates in three dimensions (3D). Therefore, the scanning laser pulses perpendicular to the direction of the flight

Fig. 7.7 RGB composition (**a**) and a thermal image (**b**) of the same area obtained with a multispectral sensor provided by the manufacturer. (MicaSense, Inc., Seattle, WA, USA).

line to register the same target several times, promoting a high precision in analyzing the results.

7.4 Flight Planning

RPAs can be manually or automatically operated. To ensure better stability during operation, flights for agronomic purposes are usually performed in an automated manner. Numerous platforms for flight planning exist, such as DJI Ground Station Pro® (Shenzhen, China), DroneDeploy® (San Francisco, USA), Pix4Dcapture® (Renens, Switzerland), and PrecisionFlight® (Raleigh, USA).

One of the main characteristics of an RPA is executing different tasks with high performance, maneuverability, and less operator intervention. However, a significant challenge in using RPAs is mainly on the definition of flight parameters. Adjusting flight parameters aims, on the one hand, to ensure sufficient quality of the images obtained for the desired purpose; on the other hand, it is also necessary to optimize the flight to allow imaging of areas in reduced time.

In short, the attributes of altitude, direction, speed, and image overlap are the main parameters that must be adjusted in a flight (Fig. 7.8). It is necessary to have in mind the purpose of the operation to make the flight planning since each operation needs specific flight configurations.

High altitude flights enable to significantly decrease the flight time to cover an area but provide images with low detail (larger pixels). Concerning altitude, an important point to note is that the RPA altitude is determined from the takeoff

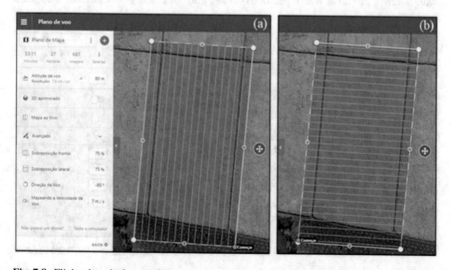

Fig. 7.8 Flight plan platform and its parameters. Example showing two different flight directions, one required fewer maneuvers and consequently higher battery autonomy (**a**), the other with maneuvers in shorter strides, requiring more maneuvers and higher battery consumption (**b**)

altitude point and kept along with the flight. Thus, the actual flight altitude relative to the ground can differ significantly throughout a survey, especially when there are changes in relief within the area flown over. Therefore, the recommendation is to start flying at the higher altitude location of the site. In this way, images collected in the lower altitude regions of the area will have larger pixels, that is, lower details. However, during image processing, it is possible to fix these pixels' distortions. In some cases, where relief changes are enormous, partitioning the area into more than one flight can solve the problems.

Flight speed has similar principles to altitude, that is, the faster the flight, the shorter the time to complete the operation. Flight speed constrains image quality according to the type of camera shutter, acquisition mode, and type of RPA platform (fixed wing or multirotor). Higher speeds result in blurred images (Fig. 7.9). It is mainly due to the camera's shutter speed being slower than the speed of flight.

Images captured at the edges of the projected area can also result in blurred images due to the effect of the maneuvering done by the RPA and the low number of overlapping images. Thus, the recommendation is to fly over an area larger than the area of interest. Figure 7.8 shows a flight plan beyond the boundaries of the field of interest.

The forward and side overlap defines image overlap. While the amount of images captured as a function of time represents the forward overlap, the flight trajectory defines the overlap. The greater the overlapping of images, the greater the number of images. As a result, the more homologous points will be identified when processing the images in the mosaicking process (described in the "image processing" section), increasing image processing accuracy. However, excessively increasing the number of images requires more time to fly and image processing time. For most applications, 70–80% overlaps, both front, and side are sufficient for adequate mosaicking.

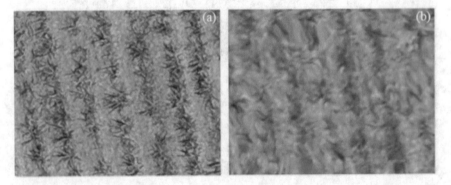

Fig. 7.9 RGB image of the exact location captured at proper flight speed (**a**) and another with blur from excessive RPA speed (**b**)

7.5   Image Processing

RPA image processing consists of techniques that allow the extraction of information from images, enabling pattern recognition and classification. However, for successful processing, the choice of RPA/camera set and flight planning settings need to match the mapping objective.

Several software programs currently exist on the market, such as Agisoft Metashape® (Agisoft, St. Petersburg, Russian) and Pix4Dmapper® (Renens, Switzerland). In addition to software installed on physical machines, cloud processing platforms also exist, such as DroneDeploy® (San Francisco, USA) and Pix4Dcloud® (cloud version of the Pix4D platform). When choosing between a web platform and software, the user needs to consider a few issues. The main advantage in processing images by software may be that the user has more control over the parameters used, such as manual settings in parameters like brightness and contrast correction, quality in filtering what soil and plant is, and adjustment of geographical location positioning. However, this type of processing demands more considerable computing resources (hardware), solved when using cloud processing. In contrast to traditional computing, cloud processing is suitable for any mobile device with an internet connection.

In general, image processing steps consist of transforming a set of projected images into a 3D structure. The first processing step is image alignment, which uses the coordinates of an image's center of perspective to group the overlapping individual images. Then the point cloud, which is the 3D representation of the terrain, is established. These points are helpful to determine 3D terrain coordinates and to find the homologous points of the images. This dense cloud of homologous points allows generating digital elevation models (DEM; its uses are discussed later) and orthomosaics.

Orthomosaics is an essential cartographic product used for several purposes, such as area, perimeter and distance measurements, vegetation indices, etc. Orthomosaics is a rectified image in the vertical, horizontal, and elevation planes (orthogonal projection) with geographic coordinates representing a natural surface. However, to have greater accuracy in the positioning of this natural surface, it is necessary to adopt georeferenced images with high accuracy, either through the survey of control points (CPs) in the field or the use of RPAs with GNSS-RTK systems.

7.6   Applications of RPAs in Agriculture

Now that we have presented the main aspects associated with the RPA/camera set and on flight planning and image processing, we will cover the leading products and applications of RPAs in agriculture: topographic survey, crop monitoring, and application of phytosanitary products, among other applications that have been appearing in the field of digital agriculture.

7.7 Topographic Survey

Topographic surveys are demanded in several fields of knowledge, not only in agriculture but also in landscape planning and civil construction. In agriculture, these data are fundamental for the correct management and conservation of the soil. Knowing terrain aspects such as altitude, slope, and associated indices allows systematizing land use such a level planting and creating terraces properly. Traditionally, these topographic surveys are carried out by classical procedures using tools such as theodolites and total stations. More recently, Global Navigation Satellite Systems (GNSS) are used, combined with techniques of differential signal correction (DGNSS), which allows reaching centimeter, or even millimeter, accuracy. However, depending on the minimum accuracy required in these traditional surveys, the work of obtaining information in the field can be expensive and time-consuming, limiting its use.

Photogrammetry can be of great help in topographic surveying. This technique is not new, as it began before the advent of airplanes through photographs taken from balloons. However, with recent advances in remote sensing technology, this technique has been perfected and is increasingly accessible. In addition to image collection by planes, today, this technique can be performed by sensors embedded in satellites and, in the specific case of this chapter, by RPAs.

In topographic surveys via RPAs, usual survey techniques (total station or GNSS) are still required but in much smaller quantities. In this case, the punctual data obtained by these technologies called control points (CPs) assist in georeferencing images, linking the coordinates of these known points with their respective position in the acquired image. The number of control points should be as high as the need for survey accuracy and or the size of the imaged terrain. In addition, the precision and accuracy in georeferencing images can also vary depending on the type of GNSS system embedded in the RPA. The most common RPAs have a GNSS system for navigation purposes, which gives them precision at a meter scale and, therefore, insufficient for higher accuracy demands (centimeters), which is the case with most topographic applications demanding greater rigor with the establishment of the CPs on the ground. Aiming to reduce this demand for fieldwork, RPA can carry GNSS of superior quality, including a real-time kinematic correction system known as RTK. Regardless, Forlani et al. (2018) state that control points are still necessary for surveys that demand high accuracy, mainly for the stability of the camera calibration parameters.

When data obtained aims at topography and or visualization in three dimensions (3D) of the terrain, the product initially obtained is called Digital Elevation Model (DEM). The DEM can lead to two other products, or better said, be named in two different ways depending on the information it provides: (1) digital surface model (DSM), which is when the product (image) contains information about the surface of the terrain, considering any object that is above the ground, as in the case of vegetation cover, and (2) digital terrain model (DTM), which provides information referring only to the ground level, that is, it does not consider what is above it,

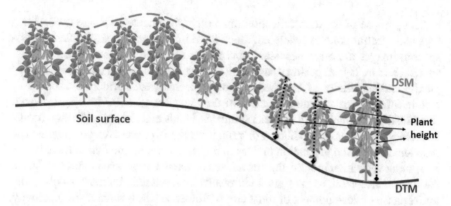

Fig. 7.10 Representation of the digital surface model (DSM; top of the canopy), digital terrain model (DTM; ground surface), and the difference between them, providing an estimate of the height of the vegetation at each location

demanding a dense point cloud filtering, so that remove the objects (dots) that are above the terrain. From the DTM, several topographic indices can be calculated, such as field elevation, slope, curvature, and water flow, among many others. Also, with these two main models (DSM and DTM), a third product can be generated by the difference in height between them, for example, a plant height or canopy volume model (Fig. 7.10).

DTM allows planning the terracing of agricultural areas and level planting and various other soil conservation practices. Furthermore, through the detailed three-dimensional reconstructions of the landscape, it is possible to estimate the susceptibility to erosion and the volumetric calculation of erosion itself. Regarding plant measurements, the 3D digital reconstruction of the crop canopy allows detecting anomalies in plant development, such as abrupt changes in its inclination, which may be related to canopy's damage due to plants falling over or pests' attack.

7.8   Crop Monitoring

One of the main focuses of using RPAs in agriculture today is monitoring crops. Through such spatialized monitoring, farmers can optimize agricultural management within the scope of precision agriculture. In this sense, we can divide the information extracted from the images into three large groups: (1) physiological assessments, (2) biophysical assessments, and (3) identification of other biological targets. Physiological evaluations seek to infer the vigor and health of the crop. At the same time, the biophysical assessments aim to obtain more direct measurements of the vegetation, such as biomass accumulation and canopy height estimation. The target identification seeks to monitor the presence of agents that reduce agricultural production (pests), such as insects, disease, and weeds.

The purpose of the data collection must drive the choice of image (or camera) type and spectral product generated. Since RGB images may be insufficient, multispectral images are often necessary, mainly for the bands located in the infrared spectral region. Hyperspectral images can also be desirable when there is high complexity and specificity of the phenomenon to be mapped. Furthermore, bands of certain infrared regions are indicative of the presence of water in leaves or soil. Individually or combined in vegetation indices, bands highlight specific target characteristics (as discussed in Chap. 4 on remote sensing), or can even be integrated as individual bands and vegetation indices in machine learning prediction models.

Among vegetation indices, the normalized difference vegetation index (NDVI) is the most widespread, as it offers good results in agronomic inferences during the initial phase of development of most crops. However, NDVI often loses sensitivity during the more advanced stages of growth due to the high leaf area index and relative homogeneity of the canopy throughout the crops. In this case, the main problem is in the red band, as its reflectance reaches a deficient level in situations with high canopy coverage, losing the ability to differentiate crop vigor, known as "saturation." Thus, it is often preferred to use the red-edge instead of the red band, resulting in the index known as NDRE. There are several other vegetation indices, each with its specific applications. A large list is at https://www.indexdatabase.de/db/i.php.

7.8.1   Physiological Assessments

This type of evaluation seeks to obtain information about the "vegetation health," that is, to infer variations in crop vigor and health through perceptible physiological changes in its spectral behavior (see Chap. 4). Such behavior, expressed by the plants based on their reflectance, can be captured by cameras carried by RPA and, through the analysis of these data, guide some decision-making. In some cases, the plant can express behavior easily identified by the visible spectrum, such as wilting, yellowing, and reduced development. However, those behaviors need sensors capable of capturing specific reflectance signals acquired from multi- or hyperspectral sensors.

With spatiotemporal information about the vigor and health of the plants, farmers can plan the site-specific management. Examples of such management are (1) targeting the application of fertilizers in variable rates based on crop vigor variations and (2) monitoring plant water stress through images to ideal irrigation on time, space, and volume (Chap. 9 addresses other issues related to precision irrigation).

Nutritional deficiency results in plant vigor reduction, usually decreasing leaf area and changes in leaf color. Thus, theoretically, the identification of such deficiencies is possible via remote sensing. However, in field situations, plants are often simultaneously susceptible to various types of stress, making it impossible to isolate the symptoms of a single problem through images. Therefore, it is necessary to be very careful when stating that nutritional deficiency can be assessed by remote sensing, especially when talking about RGB or multispectral cameras. Thus, the images can lead to problematic regions of the field, places with low crop vigor. Then, the user can go to the area to identify what is causing this vigor reduction or assume that the problem is by a given factor.

An example of this would be to assume that nematodes cause some spots that present low crop vigor when knowing that this area recurrently faces this problem; however, many other pests and factors can result in a similar symptom. An alternative used to improve the performance of image-based monitoring systems is to associate the information obtained by remote sensing with meteorological data. It is known that certain diseases occur with greater intensity under certain climatic conditions. So, if weather conditions are favorable for the occurrence of a particular disease, this information accounts for the diagnosis process. Thus, technical knowledge remains fundamental, both to interpret the symptoms in the field and understand what other information is helpful to assist in such diagnosis and, consequently, support the site-specific management.

With a focus on variable-rate fertilization, nitrogen fertilization (N) has the most significant potential to be aided by remote sensing. Soil analysis in the laboratory does not provide reliable information about the nitrogen availability for plants. Furthermore, N deficiency causes yellowing of old leaves, resulting in a spectral shift detectable by multispectral cameras and even RGB. Some approaches based on the spectral information may result in N's prescription on variable rates. Perhaps one of the most potential uses is the N-rich strips approach, meaning cultivate strips ensuring a proper N supply. In this case, a small portion of the crop receives the recommended rate of N fertilizer right at the beginning of the crop cycle. In contrast, the rest of the area usually receives a fraction of it (seeding-base fertilization), serving as a reference in the diagnosis of N demand for the rest of the field (Arnal and Raun 2018). At the ideal time for top-dressing N fertilization, spectral response collected to quantifies how much the crop is responding to the N applied, that is, how much fertilization is increasing the vigor of the crop. Thus, quantifying the variability in terms of N status and plant growth during the crop can determine whether fertilizer application in variable rates is worth implementing in the crop (Argento et al. 2020).

A wide range of studies assessed water stress in crops, mainly motivated by the flexibility of RPA use. Thus, RPAs can be an attractive alternative for irrigation planning based on monitoring the temperature of the plant's canopy through thermal cameras or the turgor of the leaves measured in some features of the electromagnetic spectrum. Similarly, instead of measuring the crop canopy, images data can infer the soil characteristics, such as surface temperature and soil moisture. Furthermore, imaging with multi- or hyperspectral sensors can help identify susceptible and tolerant plant genotypes to water stress.

7.8.2   Biophysical Assessments

Information on the variability of crop biophysical attributes, such as growth stage, canopy volume, tree canopy structure, biomass accumulation throughout the cycle, and general crop conditions, is beneficial for farmers to monitor their crops and plan for their crops further management.

The height and structure of the plant canopy is an important variable to assess the general condition of crops, obtain evidence of biotic or abiotic stress, or assist in the estimates of aboveground biomass and yield, among other information. Fertilization management and phytosanitary control are significant examples of such data uses. Furthermore, the relationship between plant height and stem (or stalk) height can be used in models to assess water stress and crop sensitivity to falling over; therefore, such an estimate is desirable in field phenotyping experiments (Walter et al. 2019).

Agricultural 3D modeling can use different sensors and image processing techniques. Within the set of sensors that can be RPA-board for agricultural applications are ranging sensors, whose data generates the point clouds and 3D models. Among them, two technologies stand out: Light Detection and Ranging (LiDAR) technology, often called laser scanning and used in several areas of knowledge, such as civil engineering and geography, and RGB-D cameras. The latter is considered a relatively inexpensive and efficient way to generate depth data, as it is a camera that captures an extra value in each RGB pixel. This value indicates the distance from the sensor to a given point in the image. A simpler and more accessible alternative to the mentioned technologies is processing high-resolution images associated with the structure from motion algorithms (SfM). It allows the generation of dense clouds of points and 3D models through the same camera used for other purposes (e.g., RGB), such as physiological assessments. A key to this method is calculating camera position, orientation, and scene geometry from the set of superimposed images. The whole three-dimensional structure of the scene results from the high overlap between images viewed from a wide range of possible positions using overlaps in the order of 90%. The objective of digital elevation models (DEM; described above when addressing topographic surveys) is to obtain the average height of the crop canopy or, in the case of perennial crops, individual plant height and crown volume (as the example in Fig. 7.10) regardless of the method of obtaining the 3D images.

An important point about this 3D modeling is that the accuracy of the measurements depends on the adequate collection and modeling of data, mainly on the digital terrain model (DTM) that will be the calculation basis. Therefore, the user can evaluate the situation depending on factors such as the crop, terrain coverage, availability of previous data, availability of control points, and the possibility of field campaigns. Thus, the most common methods are as follows:

1. Create the DTM from the interpolation of the control points values or through points collected by high-accuracy GNSS coupled to agricultural machinery, as in crop yield monitor.
2. Obtain the DTM by segmentation of the DEM pixels collected during the crop development, separating vegetation, and soil pixel.

3. Obtain the DTM from RPA flights carried out before sowing or after harvesting when there is no vegetation above the ground; that is, the DTM will be equal to the DSM.

Aboveground biomass is a particularly problematic feature to measure in the field due to the laborious and destructive methods required to assess it. The nondestructive biomass estimation extensively investigated using 2D images seeks to relate biomass data to vegetation indices. However, the possibility of relating biomass with height information measured by 3D models has great potential, especially when merging this 3D model with color images and vegetation indices (Viljanen and Honkavaara 2018). In addition, this 3D approach associated with physiological assessments through spectral behavior can provide information on canopy volume and crop health, guiding the use of inputs. For example, this approach can drive high rates of N fertilizer where the vegetation volume is high and the presence of yellowing leaves, which indicates a high chance of N deficiency occurring. At the same time, it can also guide N rate reduction when biomass is low and with high chlorophyll content (very green plants), which indicates that the limiting factor for plant development probability is not nitrogen. Pasture management is another example of identifying issues such as animal support capacity and the need for pasture reform and direct the application of pesticides and fertilizers with their recommended rate based on the canopy volume, usually proportional to the aerial volume plants.

The temporal monitoring of crops is also an exciting application in addition to this static application over time, that is, inferences drawn based on images collected at a given moment in the crop cycle. The so-called 4D point clouds are a group of several 3D images collected throughout the season that allow modeling plant growth and identifying abrupt changes in their development. With this, it is possible to develop warning systems when a particular region of the crop presents a different temporal behavior than expected, such as when pests attack the crop or when the plants fall over. In addition, it is used in the field of machine learning prediction, in which the algorithm "learns" how the crop develops and, combined with other variables such as accumulated temperature and precipitation, allows more adjusted yield predictions.

Another great application of RPAs in agriculture is the fast and efficient counting of plants and fruits. Through different image segmentation procedures, it is possible to identify the targets of interest. Thus, it is possible to determine plant and soil to carry out a seedling emergence count, for example. Similarly, it is possible to identify orange fruit in the tree crowns, allowing quantification. However, despite being a promising technique, we warn of several challenges: the need for greater spatial and spectral resolution to differentiate specific targets, establish the correct time for image collection, and the need for processing capacity and creation of robust data. Thus, this type of identification is one of the research fields that has received the most attention for research and development of computer vision techniques in agriculture.

7.8.3   Identification of Biological Targets

Another main focus of computer vision techniques using machine learning is identifying agents that reduce agricultural production (pests). Such pest monitoring is a routine to promote proper phytosanitary treatment and ensure crop yield. In this regard, it includes mainly weeds, insects, and phytopathogenic agents (plant diseases). The on-site identification methods of these agents are usually manual and require several samplings to perform a representative field evaluation. Thus, the use of RPAs for such diagnoses seeks to automate and ensure greater representativeness of the assessments, given their relatively high operational performance and high spatial resolution, filling a gap in satellite monitoring. The challenges of this operation are to identify characteristics that directly or indirectly differentiate the targets, classify them correctly, and promote automated learning by computational means.

Proper characterizations come from the information contained in good-quality images. The high spatial resolution of RPA images potentially increases the accuracy of the investigation, allowing the identification of the target of interest or the damage caused by it in the plants. However, a significant limitation of this technique is that quality orthomosaics with high spatial resolution require many images, resulting in longer flights and image post-processing with high computational cost, which still makes many applications unfeasible. The spectral quality of the embedded sensors also contributes to the target identification results, so narrower bands of multispectral sensors can be more interesting than spatial resolution, for example, in identifying weeds. In this aspect, the use of thermal bands that have a coarser spatial resolution is less frequent. However, when there is integration with other sensors, it can improve the identification of diseases and pests, even at early stages (Modica et al. 2020), given the sensitivity to physiological changes in plants.

Besides the phytosanitary control representing a significant impact on production costs, there is a strong demand for mapping insect pests, diseases, and weeds due to the environmental appeal in reducing the use of pesticides. Thus, much research developed technology associated with this issue, including commercial solutions, mainly for weed mapping and control.

7.9   Application of Phytosanitary Products via RPAs

The increase in data acquisition about crops through the different technologies associated with digital agriculture has allowed the identification of intervention needs in crops with more excellent quality, speed, and accuracy. Thus, in addition to assisting in data collection (as has been presented so far in this text), RPAs have the potential to carry out site-specific interventions within the scope of precision agriculture. In this sense, spraying of agricultural pesticides stands out since traditional spraying systems, whether manual, tractor, or aerial, do not always meet the peculiar needs of the production system. In Asia, especially China, South Korea, and Japan, there is a history of increasing technical-scientific development of spraying with RPAs, fostered by local characteristics such as small production units established in sloping areas. Currently, the RPA sprayer is a handy tool in situations not addressed, entirely or partially, by traditional equipment. Vertical takeoff, ability to maneuver in a small space, automated and geolocated operation, quick access to points of interest, and difficulty to access without causing damage to the crop are some of the exciting features that make RPAs suitable for agricultural spraying.

The service of spraying often employs the vertical takeoff RPAs with a single-engine or multi-rotor. Most of the RPAs are electricity-powered; however, there is larger equipment and load capacity with combustion engines and hybrid formats. The load capacity and the liquid flow rate applied by an RPA limit its general application to replace traditional spraying machinery. The solution tank capacity of RPAs is usually between 5 and 20 l, but there are models up to 30 l. RPAs with greater load capacities do not seem to be a technological limitation for the development but rather a consideration of manufacturers. More stringent regional legislations restrict the operation of RPAs due to the load capacity and the mandatory specific training of operators for the operation of the aircraft and the handling of chemical and biological products. As a consequence, application flows decreased to adapt to the legal processes and technical needs.

Sprayer RPAs have usage, although experimentally in most situations, in varied crops. However, the efficiency of treatments performed with this type of equipment still raises doubts. However, research has shown results consistent with other forms of spraying (Wang et al. 2019; Xiao et al. 2020), even though intermediate parameters for evaluating the quality of spraying, such as deposition uniformity, have remained with performance below other application modalities in some situations (Wang et al. 2019). On the other hand, the control efficiency achieved can be even more significant due to the solution's ability to reach lower regions of the plant due to the air vortex (air current field formed toward vegetation) created by aircraft propellers. It depends on the pest or disease targeted, location in the plant, and how the active ingredient is applied.

To obtain efficient spraying, it is necessary to correctly choose and adjust the spraying system depending on the environmental conditions and the characteristics of the targets. The proper choice of spray nozzle and flow rate determines the correct coverage of the target and the consequent efficiency of the phytosanitary treatment and reducing drift problems. The rotation of the engines changes with RPA spread and alters the vortex and consequently may impact how the pesticide reaches and or penetrates the plant canopy. The load variation during RPA flight, due to the emptying of the tank, directly reflects on the control of the rotation of the motors, causing variation in the airflow and consequent change in the distribution of droplets and the depth of penetration of the applied solution (Berner and Chojnacki 2017). The stronger the air vortex, the greater the deposition in the flight path, while with the very weak vortex, the deposition is subject to the direction and force of the wind in the environment (Guo et al. 2019). Therefore, the choice of height, speed, and direction of flight is essential for the satisfactory result of the pesticide application.

In short, there is a notable technological development regarding spraying with RPAs from 2015 onward. The diversity of agricultural production systems and the different potential demands to be addressed by RPA sprayer challenge the constant development and adaptation of these spraying systems, requiring further investigations in the search for results that make the activity functional and efficient, at the same time as safe for people and the environment.

7.10   Other Applications

Other diverse applications of RPAs in agriculture happened in recent years, and several of them are already applied infield practice. We chose to list two of them: RPA for input distribution and RPA for data transferring. There are also straightforward applications of RPAs in livestock, but they are not listed here, as remote sensing applications in cattle are in Chap. 10.

Studies aim to use RPAs as a tool that acts to distribute inputs, such as fertilizers and seeds (Weng et al. 2006). This application seems feasible in particular situations, and perhaps, there are more convenient options in the field of agricultural robotics (see Chap. 8). However, the distribution of natural enemies more efficiently in the field than would be carried out via aircraft or ground equipment is gaining users on a large scale. It has been the focus of product development by startups. The RPA transports the natural enemies encapsulated in some dispenser released into the field in a previously planned way. Such operation allows reducing costs compared to manual or aircraft applications. It provides application at the desired time, regardless of the crop size, limiting to land-based applicators. An emblematic example is the release of the *Trichogramma galloi* wasp to combat the sugarcane borer (*Diatraea saccharalis*) in commercial sugarcane fields.

Another application proposed is the use of RPA as a data transfer center. With the intensification of digitization in agriculture, more intensive use of sensors and machines is collecting data throughout time and space within the fields. Thus, connectivity in agricultural areas is growing fast, as discussed in Chaps. 11 and 12. In some developing countries, digital agriculture technologies often come up against this issue. Thus, seeking a solution to this problem, there are proposals for using RPA as a kind of "antenna" for data transfer. Hypothetically, RPAs would make regular flights in the area, daily or even several times a day, individually or as a swarm (network of RPAs that communicate to cover a larger area), and would receive data from equipment and machinery in the field. They could then store the data to be manually downloaded by the operator, instantly transfer the data to other stations installed across the farm, or transfer the data to the cloud by finding regions on the farm with connectivity. The purpose of this approach would be to enable the adoption of intelligent systems in places where the infrastructure is precarious or where the size and format of the agricultural lands do not make the installation of data transmission networks economically viable (Razaak et al. 2019). Still, in a more advanced (future) scenario, the RPA could receive this data, process them in real time, and are already acting in some way in the field, either applying some input or activating other machines to do so.

The fact is that the RPAs have the potential to help in the site-specific management of crops. What is needed now is for the user/technician to be clear about the potential of the technology, its limitations, and where they can act in a complementary way to better use equipment and available resources.

Didactic Resources of Chapter 7: Agricultural Drones' Application

(I) Abbreviations/Definitions.

- Beyond Visual Line of Sight (BVLOS): An operation way in which the pilot cannot see the RPA in visual range, even by using an Assistant Observer.
- Digital Surface Models (DSM): A digital model that represents information about the terrain surface considering any object above the ground, for example, the vegetation cover.
- Digital Terrain Models (DTM): A digital model that only provides information concerning the ground level; that is, it does not consider what is above it. It removes all objects (dots) above the terrain through a dense point cloud filtering process.
- Extended Visual Line of Sight (EVLOS): An operation way in which the pilot sees the RPA in-flight by using lenses or other long-range viewing devices.
- Global Navigation Satellite Systems (GNSS): A navigation system based on the information transmitted by satellite constellations. Devices with onboard GNSS receivers are capable of having geo-positioning information from the satellite segment.
- Ground Control Point (GCP): A visible points located on the ground with known geographic coordinates. It is useful for image ortho-correction when aerial platforms do not have an accurate GNSS receiver, such as the RTK-GNSS receiver.
- Light Detection and Ranging (LiDAR): A device that detects the position or motion of objects and operates similarly to a radar but uses laser radiation rather than microwaves.
- Normalized difference vegetation index (NDVI): A vegetation index calculated from the spectral bands NIR and red, used to assess vegetation status.
- Near-infrared region (NIR): The region of the electromagnetic spectrum between 700 and 1300 nm, which is very useful for crop monitoring.
- Red, Green, and Blue (RGB): The three bands from the electromagnetic spectrum between 400 and 700 nm, representing the visible region. They are very present in most sensors.
- Remote piloted aircraft (RPA): Synonymous with the drone.
- Structure from motion (SfM): A photogrammetric imaging technique for estimating three-dimensional structures from a set of projected two-dimensional images. Homologous information from overlapping images is

identified by feature matching algorithms so that clustering of the images occurs.

- Unmanned aircraft (UA): Synonymous of RPA.
- Unmanned aerial vehicle (UAV): Synonymous of RPA.
- Visual Line of Sight (VLOS): An operation way in which the pilot has visual contact of the RPA without assistance from an observer, use of lenses, or other long-range viewing devices.

(II) **Take Home Message/Key Points.**

- The digitalization of agriculture has opened the way for RPA applications.
- RPAs provide knowledge to advance decision-making in the field.
- RPAs prove useful for remotely capturing spatial-temporal variability with fine accuracy.
- RPAs came to assist in the site-specific management of crops such as spraying and inputs release at commercial scale.

(III) **Terms for the Index.**

- RPA legislation.
- Sensors.
- Flight planning.
- Image processing.
- Topographic survey.
- Crop monitoring.
- Physiological assessments.
- Biophysical assessments.
- RPA spraying.
- Remote sensing.

References

Argento F, Anken T, Abt F, Vogelsanger R, Walter A, Liebish F (2020) Site-specific nitrogen management in winter wheat supported by low-altitude remote sensing and soil data. Precis Agric 22:364–386

Arnal B, Raun B (2018) Applying Nitrogen-Rich Strips. Oklahoma Cooperative Extension Service, CR227. Available at: https://extension.okstate.edu/fact-sheets/applying-nitrogen-rich-strips.html. Accessed 7 July 2021

Berner B, Chojnacki J (2017) Use of drones in crop protection. Proceedings of the IX International Scientific Symposium, Lublin, Poland. pp 46–51.

Forlani G, Dall'Asta E, Diotri F, Di Cella UM, Roncella R, Santise M (2018) Quality assessment of DSMs produced from UAV flights georeferenced with on-board RTK positioning. Remote Sens 10(2):311

Guo S, Li J, Yao W, Zhan Y, Li Y, Shi Y (2019) Distribution characteristics on droplet deposition of wind field vortex formed by multi-rotor UAV. PLoS One 14(7):e0220024

Modica G, Messina G, De Luca G, Fiozzo V, Praticò S (2020) Monitoring the vegetation vigor in heterogeneous citrus and olive orchards. A multiscale object-based approach to extract trees' crowns from UAV multispectral imagery. Comput Electron Agric 175:105500

Nijland W, De Jong R, De Jong SM, Wulder MA, Bater CW, Coops NC (2014) Monitoring plant condition and phenology using infrared sensitive consumer grade digital cameras. Agric For Meteorol 184:98–106

Razaak M, Kerdegari H, Davies E, Abozariba R, Broadbent M, Mason K, Argyriou V, Remagnino P (2019) An integrated precision farming application based on 5G, UAV and deep learning technologies. Commun Comput Inf Sci 1089(2020):109–119

Viljanen N, Honkavaara E (2018) A novel machine learning method for estimating biomass of grass swards using a photogrammetric canopy height model, images and vegetation indices captured by a drone. Agriculture 8:70

Walter J, Edwards J, Cai J, McDonald G, Miklavcic SJ, Kuchel H (2019) High-throughput field imaging and basic image analysis in a wheat breeding programme. Front Plant Sci 10:1–12

Wang G, Lan Y, Yuan H, Qi H, Chen P, Ouyang F, Han Y (2019) Comparison of spray deposition, control efficacy on wheat aphids and working efficiency in the wheat field of the unmanned aerial vehicle with boom sprayer and two conventional knapsack sprayers. Appl Sci (Switzerland) 9(2):1–17

Weng LX, Deng H, Xu J, Li Q, Wang L, Jiang Z, Zhang HB, Li Q, Zhang L (2006) Regeneration of sugarcane elite breeding lines and engineering of stem borer resistance. Pest Manag Sci 62(2):178–187

Xiao Q, Du R, Yang L, Han X, Zhao S, Zhang G, Fu W, Wang G, Lan Y (2020) Comparison of droplet deposition control efficacy on *phytophthora capsica* and *aphids* in the processing pepper field of the unmanned aerial vehicle and knapsack sprayer. Agronomy 10(2):215

Zhang N, Su X, Zhang X, Yao X, Cheng T, Zhu Y, Cao W, Tian Y (2020) Monitoring daily variation of leaf layer photosynthesis in rice using UAV-based multispectral imagery and a light response curve model. Agric For Meteorol 291:108098

Chapter 8
Sensors and Actuators

Daniel Marçal de Queiroz, Domingos Sárvio M. Valente, and Andre Luiz de Freitas Coelho

8.1   Introduction

Agricultural machinery used in precision agriculture and digital agriculture are equipped with a wide range of sensors and actuators. To work with precision agriculture and digital agriculture, it is necessary to understand the basic principles of the sensors and actuators that are present in these machines.

Sensors are devices that interact with the physical environment and are capable of producing a signal that is related to the variable to be measured. In most sensors, the signal produced is electric, but another type of signal can be generated. For example, in the mercury thermometer, the signal is the level that the fluid reaches within the graduated column. When this signal is electric, it has several advantages, such as ease of reading, of using to control a process, of recording, and of transmitting the generated data.

Agricultural machines for precision agriculture and digital agriculture make use of various sensors. Two groups of sensor application in agricultural machinery are the monitoring of an operation and the control of an operation. The first group includes the sensors used in the yield monitor of the harvester. The second group contains sensors that are used to monitor the variables that will be used to determine the doses to be applied or to control the metering mechanisms. Depending on the type of machine and the operation to be performed, different types of sensors are employed. Examples include sensors of displacement speed, angular velocity, level, position, mass flow, volumetric flow, temperature, water content, pressure, force, torque, soil apparent electrical conductivity, soil organic matter, and pH, among others. Each sensor has its usage characteristics that depend on how it was designed and the material and methods employed in its construction.

D. Marçal de Queiroz (✉) · D. S. M. Valente · A. L. de Freitas Coelho
Department of Agricultural Engineering, Universidade Federal de Viçosa, Viçosa, MG, Brazil
e-mail: queiroz@ufv.br; valente@ufv.br; andre.coelho@ufv.br

© The Author(s), under exclusive license to Springer Nature
Switzerland AG 2022
D. Marçal de Queiroz et al. (eds.), *Digital Agriculture*,
https://doi.org/10.1007/978-3-031-14533-9_8

While sensors react to changes in the medium and produce a signal, actuators are devices that operate by doing the opposite, that is, from the current received signal about the medium, moving or controlling something from the surrounding medium. In variable-rate application machines, for example, actuators are used to automatically vary the rate of agricultural inputs to be applied. These actuators perform their function according to the signal received from the machine control system.

Many applications use intelligent sensors and actuators, and devices with information processing capacity (microprocessors) often are integrated with conventional sensors or actuators. In sensors, for example, microprocessors are used to receive a signal and process it in order to facilitate the subsequent steps. In intelligent actuators, microprocessors can be used to receive the signal from the control system and convert it into the form requested by the actuator. Examples of intelligent sensors and actuators include those capable of sending or receiving data using communication protocols such as CANbus (Controller Area Network), I²C (Inter-Integrated Circuit), SPI (Serial Peripheral Interface), Bluetooth, Wi-Fi, and Zigbee, among others.

As presented, sensors and actuators are widely used in digital agriculture. They equip monitoring systems and integrate automated systems for distribution of inputs, which are generally called variable-rate application machines. This chapter presents sensors used for the mapping of soil attributes, followed by sensors used for plants, concluding with those used in yield mapping systems. In addition to the sensors, the types of actuators used in the machines are also presented.

8.2   Mapping of Soil Attributes

Mapping soil attributes is an important step in digital agriculture, since crop yield is directly influenced by the amount of nutrients available in the soil. Fertilizers and correctives make up a substantial part of the production cost of crops. Then, the rates of these inputs must be determined carefully if the objective is to optimize the production system. Agricultural soils are not uniform, and there is no defined pattern of spatial variability of soil attributes. In addition, each attribute can show different behaviors of spatial variability.

Mapping soil attributes is not an easy task. The cost of obtaining soil attribute maps and the accuracy of the generated maps are conflicting. To obtain maps with a low level of error, it is necessary to adopt a grid with a high density of collection points. However, the cost increases proportionally to the number of points.

There are several methods to characterize the spatial variability of chemical, physical, and physical-chemical attributes of soils. Characterization can be made using sensors or by collecting soil samples with subsequent analysis in the laboratory. As for the definition of measurement or sampling points, grid sampling or directed sampling may be used.

8.2.1 Soil Sampling

Grid soil sampling is widely used to characterize the spatial variability of soil attributes and generate fertilizer rate maps. It is common to adopt grids with one composite sampling point every 3–5 ha, even adopting up to one point every 10 ha. Each sampled point consists of several single soil samples, collected within the area considered. The problem is that these grids are often not dense enough to accurately represent the spatial variability of soil attributes. To generate maps by interpolation, Fergunson and Hergert (2009) recommend that soil sampling for precision agriculture purposes be done with one point every 0.40 ha (Fig. 8.1), collecting five single samples within a radius of up to 3 m around the point, to generate a representative composite sample of the point. For soils with lower spatial variability, it is recommended to use up to one point per hectare. According to these authors, the use of these grids makes it possible to obtain maps that are valid for several years: 10–20 years for the attributes organic matter and cation exchange capacity, 5–10 years for pH, and 4–5 years for phosphorus, potassium, and zinc.

In the directed sampling, previous information should be obtained in the field and then used to divide the area into management zones. For each zone, one composite sample is collected, which characterizes its soil condition. Each zone can be delimited, for example, from the yield maps of several harvests. The area can be divided into areas of high yield, medium yield, and low yield. In each of these areas, one soil sample is collected to characterize it, and the rate of fertilizers and correctives to be applied is defined. Other information that can be used in the delimitation of these zones are aerial images (obtained by satellites, aircraft, or unmanned aerial vehicles), maps of soil apparent electrical conductivity, digital elevation models, and maps of soil types. Therefore, despite having a cost well below that of grid sampling, directed sampling is only possible if there is prior information about the area.

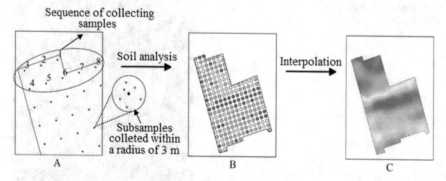

Fig. 8.1 Soil sampling grid scheme for precision agriculture purposes (**a**), identification of collection points (**b**), result of each point for a given soil attribute and (**c**) map of the interpolated attribute. (Source: Authors of this chapter)

To increase the operational capacity of soil sampling, it is very common to use mechanized systems. Generally, these systems use an all-terrain vehicle, which reduces the journey time between collection points, and a mechanical system for sampling. The sampler can be driven by hydraulic motor, electric motor, or internal combustion engine. The vehicle is equipped with global navigation satellite system (GNSS) for navigation and georeferencing of sampling points. Figure 8.2 shows the Solo Drill system for soil sampling with threaded sampler, developed by Falker company.

8.2.2 Sensors for Mapping Soil Attributes

Due to the high cost of the process of collecting and analyzing soil samples, the development of sensors to characterize the spatial variability of soil attributes has been sought. Adamchuk et al. (2004) present the different sensors that have been used to characterize the soil, classifying them into the following types:

- Electrical and electromagnetic sensors: Measure resistivity or electrical conductivity, capacitance or inductance, variables that are associated with soil composition.
- Optical and radiometric sensors use electromagnetic waves to detect the level of energy absorbed or reflected by soil particles: These variables are usually associated with the water content and organic matter of the soil.
- Mechanical sensors: Measure the force associated with the action of a tool on the soil, a variable that is usually associated with the physical attributes of the soil.

Fig. 8.2 Solo Drill soil sampling system from Falker. (Source: Authors of this chapter)

- Acoustic sensors: Measure the amount of sound produced by the interaction of a tool with the soil, variable that is also usually associated with the physical attributes of the soil.
- Pneumatic sensors: Measure the ease with which air can be injected into the soil, variable that is usually associated with soil porosity.
- Electrochemical sensors: Use ion-selective membranes to produce an output voltage in response to the activity of certain ions (H^+, K^+, NO^{3-}, Na^+, etc.). These sensors are used to determine the chemical properties of the soil.

Sensor response is typically not associated with a single soil attribute. In addition, each type and condition of soil may show distinct behaviors. Despite these problems, various types of sensors have been used to map soils. This is due to the speed of the data collection process and the possibility of performing the measurement using a much larger number of points per unit of area.

One of the sensors that have been widely used to characterize the spatial variability of the soil is the soil apparent electrical conductivity sensor. In fact, this type of sensor was developed in the early twentieth century and, for a long time, has been used to identify areas with soil salinity problems. With the development of precision agriculture, it began to be used to identify areas that have similar soil characteristics.

Soil apparent electrical conductivity is associated with the ease with which the soil conducts electricity, which in turn is associated with the interaction between the different physical and chemical components of the soil. There are basically two methods for determining soil apparent electrical conductivity: one based on the determination of electrical resistivity and the other based on electromagnetic induction.

The method of electrical resistivity is based on the introduction of four electrodes equally spaced into the soil surface. An electric current is applied to the external electrodes, and the potential difference is measured in the internal electrodes (Fig. 8.3). This array is called **Wenner Matrix**. The method was developed around 1920 by Conrad Schlumberger in France and Frank Wenner in the United States to evaluate soil apparent electrical conductivity (Corwin and Lesch 2003). A

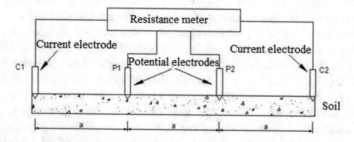

Fig. 8.3 Electrical resistivity method for determining soil apparent electrical conductivity with four electrodes: two current electrodes (C1 and C2) and two potential electrodes (P1 and P2). (Source: Adapted from Corwin and Lesch 2003)

higher number of electrodes and with other arrangement and spacing configurations can be adopted, depending on the depth of the soil profile to be measured.

The apparent electrical resistivity of the soil for the array presented in Fig. 8.3 (Wenner matrix) is determined by Eq. 8.1. Soil apparent electrical conductivity is the inverse of resistivity and is determined by Eq. 8.2.

$$\rho = \frac{2\pi a \Delta V}{i} \qquad (8.1)$$

where:

ρ = soil apparent resistivity, Ωm;
a = spacing between electrodes, m;
ΔV = potential difference measured, V;
I = electric current applied to external electrodes, A.

$$EC_a = \frac{1}{\rho} \qquad (8.2)$$

where:

EC_a = soil apparent electrical conductivity, Sm^{-1}.

The term apparent electrical conductivity is used because the electric current can flow through the soil in three ways, and what is measured is a value that represents the three forms of electricity conduction. These phases are as follows: (a) in the liquid phase, through the solids dissolved in the water contained in the large soil pores; (b) in the solid-liquid phase, through exchangeable cations associated with clay minerals; and (c) in the solid phase, through solid particles that are in contact with each other. Due to these three phases, electrical conductivity can be influenced by various physical and chemical properties of the soil, such as salinity, degree of saturation, moisture, and density. Moreover, mainly through exchangeable cations, the electrical conductivity is influenced by the content and type of clay, cation exchange capacity, and organic matter (Corwin and Lesch 2005b).

Another way to measure soil apparent electrical conductivity is by electromagnetic induction. This type of sensor was also initially used to measure soil salinity and then came to be used in precision agriculture. The sensor works close to the soil, but without direct contact, using signal at frequencies from 0.40 to 40 kHz. The reduction in signal intensity is used to determine soil apparent electrical conductivity (Corwin et al. 2008). The system consists of two coils, a transmitter and a receiver (Fig. 8.4). The transmitter is located at one end of the device that induces Foucault current loops in the soil. The magnitude of the loops generated is directly proportional to the apparent electrical conductivity of the soil in the proximity of the loop. Each current loop generates a secondary electromagnetic field, which is proportional to the current flow in the loop. The two signal is detected at the receiver

Fig. 8.4 Principle of operation of the electromagnetic induction sensor. (Source: Corwin and Lesch 2003)

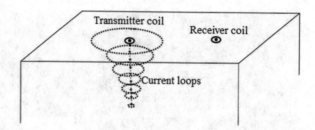

coil, one related to the primary electromagnetic field and the other related to the secondary electromagnetic field. The apparent electrical conductivity is proportional to the ratio between the secondary and the primary magnetic field intensities.

The apparent electrical conductivity has high correlation with soil water content. Hence, there arises the question about under which soil moisture condition electrical conductivity measurements should be performed. The electromagnetic induction sensor has no restrictions with respect to soil water content, as the sensor does not touch the soil during determinations. Conversely, the sensor based on electrical resistivity has problems when the soil is very dry, since the physical contact between the electrodes and the soil becomes more difficult, compromising the passage of current from one electrode to another (Corwin and Lesch 2005a). Studies have shown that the sensor should be used with moisture close to field capacity. Some manufacturers of this sensor recommend moistures above 20%.

Apparent electrical conductivity maps are used to characterize the spatial variability of soil conditions. The idea is that, from the apparent electrical conductivity, it possible to divide the area into zones that have similar soil conditions. The important point to be verified is whether, with different soil moisture conditions, the spatial variability pattern changes. If the variability pattern changes, then electrical conductivity could not be used to characterize soil variability, or a soil water content condition would have to be defined so that the tests could be performed. However, research studies have confirmed that the value of soil apparent electrical conductivity changes, but the variability pattern remains the same when the determinations are carried out under different soil moisture conditions.

Other sensors have been developed for soil characterization, and one of them is the pH sensor. There is more than one principle that can be used by sensors to measure soil pH, and one of those commercially adopted is that which uses selective ions in antimony electrodes. Another type of sensor used for soil mapping is an organic matter sensor based on soil reflectance at certain wavelengths. This type of sensor assumes that the organic matter content modifies the spectral signature of the soil.

The company Veris Technology is one of the manufacturers of soil apparent electrical conductivity sensors. One of the common models manufactured by the company is the Veris 3100, which has recently been replaced by the Veris 3150 model. Veris 3100 was developed to be pulled by a vehicle and is equipped with a set of six electrodes that allow for simultaneous mapping of soil apparent electrical

Fig. 8.5 Veris Technology's U3 platform, which integrates sensors of soil apparent electrical conductivity, pH, and reflectance. (Source: Authors of this chapter)

conductivity at two depths, 0–30 cm and 0–90 cm. Subsequently, the company incorporated other sensors by launching various models of equipment. Figure 8.5 presents the U3 platform, composed of a soil apparent electrical conductivity sensor with four electrodes, arranged for measurement at 60 cm depth, a pH sensor, and a sensor to determine the subsurface reflectance of the soil. Once calibrated, the reflectance sensor can be used to estimate the soil organic matter content.

An alternative to soil sensors pulled by a tractor is the portable ones. They can be used on mountainous areas where vehicles cannot be used, and their lower cost allows the soil mapping of small areas. One example of portable soil sensor is the one developed by Queiroz et al. (2020) and shown in Fig. 8.6. The sensor was developed using a BeagleBone Black single board computer. A graphical user interface (GUI) was created to help the user to do the data acquisition. The sensor works with an alternate current and allows the user to choose in the interface what frequency of the signal to be applied to soil. A GNSS module is used for georeferencing the collected data. Maps of the soil apparent electrical conductivity (ECa) and the estimated ECa standard deviation generated by ordinary krigging are available to the user in the GUI.

Fig. 8.6 Portable soil apparent electrical conductivity sensor. (Source: Queiroz et al. 2020)

8.3 Mapping of Plant Attributes

Mapping of attributes related to crop development is important information for the management of systems that use precision agriculture and digital agriculture. This information can be used to define the application of fertilizers, for combating pests and diseases and in weed control.

Mapping can be performed by visual inspection or using sensors. When the inspection is done visually, a technician or a team of technicians can go to the field and use a GNSS device to delimit the areas that need intervention and generate the maps of application of inputs. When using a sensor, these devices can be manually operated by field technicians or coupled to ground vehicles, unmanned aerial vehicles, aircraft, or satellites.

The most efficient way to map plant attributes is through systems that measure the reflectance of the plants. There are basically two groups, passive ones, which use electromagnetic radiation emitted by the sun, and active ones, which are equipped with electromagnetic radiation source.

Electromagnetic energy, when reaching the plant, can be absorbed, transmitted, or reflected by the leaves. Plant behavior in relation to electromagnetic energy is known as spectral response. The spectral response of a canopy resembles that of a healthy green leaf (Fig. 8.7), which can be analyzed in three distinct regions of energy: visible region (400–700 nm), near-infrared region (700–1300 nm), and shortwave infrared region (1300–2600 nm).

Fig. 8.7 Spectral response typical of a healthy green leaf. (Source: Novo 2010)

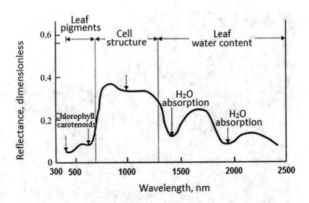

In the visible region, the reflectance is relatively low, due to the strong absorption of radiation by the optically active pigments of the leaf, showing two minimum sites of reflectance, which correspond to absorption peaks, close to 480 nm and 680 nm. Between these two minimums, there is a peak of reflectance, corresponding to the green wavelength (around 550 nm), which explains the green color of the plants. Between the wavelengths of 700 nm and 1300 nm (near-infrared region), there is high reflectance related to the cell structure of the leaves. This high reflectance is important for the leaf to maintain equilibrium in the energy balance and not over-heat, thus avoiding its destruction. Two other points of higher absorption occur near 1400 nm and 1900 nm, in the shortwave infrared region, due to the presence of water in the leaf (Moreira 2011).

The spectral response of the leaf and canopy of a given crop may vary according to a reasonable number of factors, such as the leaf area index (LAI), nutritional status of the plant, and presence of diseases, weeds, and pests. This variation in spectral response has been identified by sensors and instruments and used for decision-making and localized management of agriculture. Electronic sensors, spectroradiometers, artificial vision systems (analysis of images acquired by cameras positioned in agricultural machines), and remote sensing using multispectral and hyperspectral images are some examples of instruments used to measure spectral information in agriculture (Lee et al. 2010).

For weed control, an optical sensor can be used. As the spectral signature of the plants is different from the signature of the soil, this sensor detects whether there is a plant in the crop interrow; if so, the system sends a signal to a solenoid valve to make the real-time localized application. With the sensor positioned between the crop rows, the plant detected by it is considered weed. WeedSeeker® is a commercially available sensor that works based on this principle.

Reflectance-based sensors have been used to define the nitrogen rate to be applied. Methods for defining the rate in cereal cultivation are developed based on the normalized difference vegetation index (NDVI), given by Eq. 8.3 and proposed by Rouse et al. (1973). In this index, the difference in reflectance between the near-infrared and red bands is divided by the sum of the reflectances in these two bands to normalize the difference. The values obtained by this index vary between −1 and + 1. Bare soil usually has NDVI around 0.20. Maize crop may have NDVI up to close to 1. More vigorous plants tend to absorb more radiation in the red band, hence reflecting less energy in this band and more energy in the near-infrared band. Thus, better developed plants tend to have a higher NDVI.

$$NDVI = \frac{\rho_{NIR} - \rho_{RED}}{\rho_{NIR} + \rho_{RED}} \qquad (8.3)$$

where:

NDVI = normalized difference vegetation index;
ρ_{NIR} = reflectance in near infrared band;
ρ_{RED} = reflectance in red band.

The two most used methods for defining the amount of nitrogen to be applied in cereal cultivation are the one based on the concept of nitrogen sufficiency (Varvel et al. 2007) and the one that explores the temporal variation of NDVI and its relationship with crop yield (Solie et al. 2012). In both methods, it is necessary to cultivate a strip of the area with a nitrogen supply such that the plants are not stressed by this nutrient. This strip is called the reference strip.

The method based on the concept of nitrogen sufficiency uses the Soil-Plant Analysis Development sensor (SPAD, Minolta Camera Company, Osaka, Japan), which determines the chlorophyll index of the plant. This index is correlated with the amount of nitrogen present in the plant. Chlorophyll index is measured in the reference strip and the area in which nitrogen is to be applied. Then, the nitrogen sufficiency index is calculated, by dividing the value of the chlorophyll index of the area where nitrogen will be applied by the chlorophyll index of the reference strip. This nutrient needs to be applied if the value of the sufficiency index is lower than 0.95.

The method based on NDVI temporal variation uses the GreenSeeker sensor (N-Tech Industries, California, USA). The NDVI values of the strip where nitrogen will be applied and of the reference strip are used to determine a parameter called normalized rate. Based on the crop, expected or desired yield and normalized rate, the amount of nitrogen that should be applied is determined. This algorithm was developed based on the fact that yield varies with NDVI according to a sigmoidal function and that the parameters of this function change as plants develop. Figure 8.8 illustrates the use of GreenSeeker in a nitrogen distributor. In this system, the nitrogen rate is automatically varied as the NDVI value of the crop changes.

In addition to the vegetation index sensors cited, there are several other types in the international and national markets, basically with the same objective, for instance: N-sensor (Yara International ASA, Norway), CropXplorer (Agxtend, CNH Industrial), Crop Circle (Holland Scientific, USA), CropClorofilog (Falker Automação Agrícola Ltda., Brazil), and Flexum (Falker, Brazil).

In addition to reflectance-based sensors, there are others that have been used for crop monitoring. Crop meter is one of them and has as its principle a pendulum installed in front of a vehicle, with a sensor that measures the angle formed by the pendulum. The higher the angle, the higher the biomass density of a crop (Ehlert and Dammer 2006). These biomass density data have been used to control the application of pesticides. Another type of sensor used for this control is the

Fig. 8.8 GreenSeeker sensor for variable-rate application of nitrogen. (Source: Trimble Navigation Limited 2013)

ultrasound-based sensor; it emits small acoustic pulse and measures the response to that pulse. Its signal is used to infer about the height of plants and their amount of biomass (Dammer et al. 2015).

8.4   Yield Mapping

In precision agriculture and digital agriculture, the yield map is a highly relevant piece of information. From this map, it is possible to evaluate the areas that are producing well and those in which it is necessary to modify the management to increase profitability. In addition, tests of varieties, spacings, and rates can be conducted, and the yield map can indicate the effect of each parameter evaluated.

To obtain a yield map, it is necessary to have a system called yield monitor, which brings together a series of components capable of estimating the yield of the crop at each point in the area. Yield monitors have different characteristics, depending on the product being harvested: grains, coffee, forages, cotton, tubers. This chapter addresses only the characteristics of grain yield monitors, which are the most common.

Yield is estimated by the monitor from information produced by different sensors. It can be said that the most important sensor of a yield monitor is the one that measures the flow of grains reaching the bulk carrier tank of the harvester. By integrating the grain flow data at constant time intervals, it is possible to determine what grain mass was harvested in the interval Δt. Intervals Δt between 1 and 10 s are usually used. To obtain yield, it is necessary to determine the area where the grain mass was harvested. This area is known by multiplying the distance traveled by the machine, in the time interval, Δt, by the width of its work, and by integrating the displacement speed of the harvester over this interval. Therefore, the yield monitor needs to have a machine displacement speed sensor and another that determines the grain flow. Yield is estimated based on Eq. 8.4.

$$Y = K \frac{q \Delta t}{\Delta x L_p} \tag{8.4}$$

where:

Y = estimated crop yield, kg/ha;
q = grain flow, kg/s;
Δt = time interval between yield determinations, s;
Δx = spaced covered by the harvester since the last yield determination, m;
L_p = effectively used width of the harvester's platform, m;
K = conversion factor for the presented units, K = 10,000 in this case.

Grains are harvested at different moisture contents, so most yield monitors are equipped with a grain moisture content sensor to allow yield data to be converted into standard moisture content yield. Yield is corrected to a standard moisture content by means of Eq. 8.5.

$$Y_c = Y \frac{100 - U_c}{100 - U_s} \tag{8.5}$$

where:

Y_c = yield corrected to a standard moisture content, kg/ha;
U_c = moisture content of the product determined by the moisture sensor, % wet basis;
U_s = standard moisture content used to correct all yield data, % wet basis.

For georeferencing yield data, a global satellite navigation system, usually of the RTK type, is coupled to the yield monitor. The yield monitor captures and processes the signal from the various sensors, generating yield data every time interval. The data of yield, point position, and other values determined by the system are stored on the yield monitor and then transferred to a computer to generate the yield map. If a connectivity system is available in the field, these data can be transferred in real time to a center at the company's headquarters or to a cloud server. Figure 8.9 presents a scheme showing how various components of a yield monitor are connected.

To measure grain flow, the most commonly used sensors are volumetric and impact sensors. Volumetric sensors are usually installed in the body of the elevator that conveys the grains from the cleaning system to the bulk tank of the harvester. One of the forms adopted consists of a light emitter on one side of the conveyor and

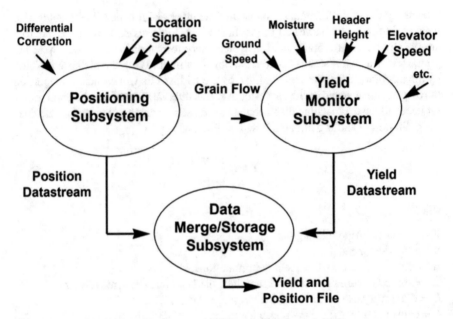

Fig. 8.9 Operating diagram of yield monitors. (Source: Casady et al. 1998)

Fig. 8.10 Operating
scheme of the volumetric
grain flow sensor. (Source:
Miu 2015)

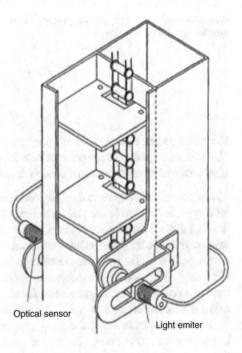

a light receiver (photosensor) on its other side (Fig. 8.10). The higher the grain flow in the conveyor, the lower the light intensity received in the photosensor, which produces an electrical signal proportional to the intensity of the light it receives. A calibration curve is developed to convert the electrical signal into grain flow. In this type of sensor, the inclination of the harvester can interfere with the performance of the system. Thus, a machine tilt sensor is used to correct the data of the grain flow.

Grain flow sensors by impact are usually installed at the conveyor outlet that takes the grains to the bulk carrier tank. The sensor measures the force applied by the grains on an impact plate (Fig. 8.11). A calibration curve is used to convert the electrical signal into grain mass flow.

The water content sensors employed in yield monitors use the principle of capacitance. One of the forms used is a parallel plate sensor installed next to the conveyor that transports the grains to the bulk carrier tank. Part of the transported grains is diverted to the sensor in order to determine the water content. Another arrangement adopted is the installation of the sensor along the conveyor thread that takes the grain from the elevator to the center of the bulk carrier tank. The parallel plates, or the body of the conveyor thread, form a capacitor, and the grains are the dielectric material. The water content of the grains varies the electrical permissiveness of the medium and, consequently, the capacitance. Electronic circuits are used to measure capacitance, and from the calibration, it is possible to correlate the capacitance with the water content of the grains.

The displacement speed of the harvester is determined using a sensor that measures the angular speed of one of its wheels. Thus, after knowing the perimeter of the wheel, the angular velocity is converted into displacement speed. Another system adopted is through a radar sensor. This sensor emits a microwave signal toward the soil and captures the signal reflected by the surface. The travel speed is proportional to the difference between the frequency of the emitted signal and the received signal. A third way to determine the displacement speed is through the GNSS receiver installed for the positioning of the harvester.

When the machine starts the maneuvering procedure, upon reaching the end of the field, there are still grains flowing through the machine components, and this flow is measured by the flow sensor. A switch connected to the platform informs the

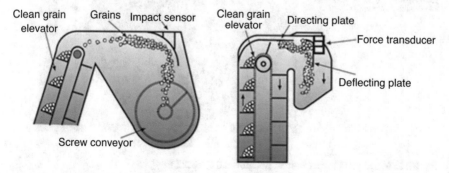

Fig. 8.11 Operating scheme of impact-plate grain flow sensors. (Source: Grisso et al. 2009)

yield monitor if the machine platform is raised or lowered in order not to generate
yield data during the maneuver operation.

Every yield monitor requires calibration, since yield is obtained indirectly from
multiple sensors. Each monitor requires its own calibration system, which is defined
by the equipment manufacturer. The calibration process involves calibrating the
sensors of the harvester's displacement speed, grain moisture content, and grain
flow. Grain flow sensor calibration is the last step in the calibration process and
involves harvesting a portion of the area until the bulk carrier tank is full. The grains
are transferred to a bulk carrier truck, and the grain mass harvested is weighed on a
scale. The obtained mass value is reported to the yield monitor, enabling it to determine the correction factors for the yield data to be generated. Depending on the
manufacturer, the recommendation of the number of weighing procedures for sensor calibration changes, and there are systems in which calibration is done with a
single point. A second possibility is to collect two points (one with high grain flow
and the other with low grain flow) and assuming that the behavior between the sensor signal and the grain flow is linear. There are still calibration systems in which
data from five or more tests performed under different flow conditions are collected.
Systems that adopt a larger number of points tend to provide better results due to the
nonlinear behavior between the signal produced by the sensor and the grain flow
passing through the sensor. This calibration process needs to be repeated whenever
there is a significant change in harvesting conditions, for example, a change in the
type of crop, variety, and condition of the product to be harvested.

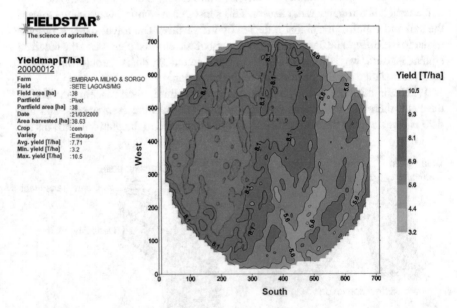

Fig. 8.12 Maize yield map. (Source: Bongiovanni et al. 2006)

Once the yield monitor is calibrated, it will be ready to use. Figure 8.12 shows a maize yield map (Bongiovanni et al. 2006). The average yield was 7710 kg/ha; however, it varied between 3200 kg/ha and 10,500 kg/ha. A qualitative analysis of the map allows identifying areas with above-average yield (West) and others with below-average yield (East). This behavior is linked to the management of this area, and no-tillage is adopted in the west region, while conventional tillage is adopted in the east region.

The calibration process is laborious and time-consuming. Therefore, many producers end up not performing calibration. Without the calibrated monitor, yield data may have a high level of error. One solution is the development of automatic calibration systems, such as Active Yield developed by John Deere. The Active Yield system incorporates three load cells at the bottom of the bulk carrier tank, and the signal produced by these load cells is used to correct the values read by the impact sensor that measures the grain flow.

8.5 Actuators in Digital Agriculture

Actuators are present in virtually all machines used in precision agriculture and digital agriculture. This component has the function of transforming one type of energy into another. In agricultural machinery, the input energy can be electric, hydraulic, or pneumatic, while the output energy is usually mechanical energy (generating movement or a force). In the three forms of input energy, the generated motion can be linear or rotating.

Among the electric actuators, the most used in agricultural machinery are electric motors and solenoid valves (or electric valves). Electric motors, which can be of alternating or continuous current, are used to trigger mechanisms that require rotational mechanical energy. Solenoid valves are used to control the flow of liquids, such as the flow of agricultural pesticides to the spray nozzles.

Hydraulic and pneumatic actuators use hydraulic oil and air, respectively. Actuators that produce rotating movements are called motors (hydraulic or pneumatic), while those that generate linear movements are called cylinders (hydraulic or pneumatic). Hydraulic actuators have the characteristics of high force and torque capacity, with lower speeds. Pneumatic actuators produce movements with higher speed but have lower force and torque capacities.

In modern agricultural machines and robots, there is an increasing tendency to use electric actuators, which allow direct integration with the electronic control system. Hydraulic and pneumatic actuators are integrated into the electronic control system through directional valves and flow regulating valves responsible for varying the direction and linear as well rotational velocity. Direct integration of the actuator into the electronic control system increases accuracy and decreases oscillations and response times. In addition, electric actuators do not require the use of hydraulic and pneumatic systems, which are complex and heavy and have a large number of components.

In modern agricultural machines and robots, there is an increasing tendency to use electric actuators, which allow direct integration with the electronic control system. The electric actuator technology has improved since the 1990s. Today, they are more reliable and affordable. Also, electric actuators can be easily integrated into the electronic control system of the machines increasing accuracy and decreasing oscillations and response times. In addition, electric actuators do not require the use of hydraulic and pneumatic systems, which are complex and heavy and have a large number of components.

Abbreviations/Definitions

- Sensors: Devices that interact with the physical environment and can produce a signal that is related to the variable to be measured.
- Actuators: Devices that make a system to develop a defined function.
- Intelligent actuators: Microprocessors can be used to receive the signal from the control system and convert it into the form requested by the actuator.
- Management zones: As areas that have the same crop yield or crop qualitative potential where a uniform dosage of input can be applied.
- Soil apparent electrical conductivity sensor (ECa): Sensor that measures the electrical conductivity of undisturbed soil.
- Electrical and electromagnetic soil sensors: Measure resistivity or electrical conductivity, capacitance, or inductance of the soil.
- Optical and radiometric soil sensors: Use electromagnetic waves to detect the level of energy absorbed or reflected by soil particles.
- Acoustic soil sensors: Measure the amount of sound produced by the interaction of a tool with the soil.
- Pneumatic soil sensors: Measure how ease air can be injected into the soil.
- Electrochemical soil sensors: Use ion-selective membranes to produce an output voltage in response to the activity of certain ions (H^+, K^+, NO^{3-}, Na^+, etc.).
- Wenner Matrix: Method measure apparent soil electrical conductivity based on the introduction of four electrodes equally spaced into the soil surface.
- Leaf Area Index (LAI): Defined as the projected area of leaves over a unit of land.
- Normalized Difference Vegetation Index (NDVI): Vegetation index calculated by the difference in reflectance between the near infrared and red bands divided by the sum of the reflectances of these two bands to normalize the difference.

Take Home Message/Key Points

- Sensors are devices that interact with the physical environment and can produce a signal that is related to the variable to be measured.
- In the digital farming, many sensors are used to determine crop yield and to monitor crop development, weather condition, soil health, and machinery operation conditions.
- Mapping soil attributes is an important step in digital agriculture since crop yield is directly influenced by the amount of nutrients available in the soil.

- Sensors that have been used to characterize soil conditions can be classified into the following types: electrical and electromagnetic sensors, optical and radiometric sensors, mechanical sensors, pneumatic sensors, electrochemical sensors.
- The most efficient way to map plant attributes is through systems that measure the electromagnetics reflectance of the plants. Electromagnetic energy, when reaching the plant, can be absorbed, transmitted, or reflected by the leaves. Plant behavior in relation to electromagnetic energy is known as spectral response. The spectral response of a canopy resembles that of a healthy green leaf which can be analyzed in three distinct regions of energy: visible region (400–700 nm), near-infrared region (700–1300 nm), and shortwave infrared region (1300–2600 nm).

References

Adamchuk VI, Hummel JW, Morgan MT, Upadhyaya SK (2004) On-the-go soil sensors for precision agriculture. Comput Electron Agric 44:71–91

Bongiovanni R, Mantovani EC, Best S, Roel A (2006) Agricultura de precisión: integrando conocimientos para una agricultura moderna y sustentable. [S.l.]: Procisur/IIC, 205 p

Casady, W. W., Pfost, D. L., Ellis, C. E., Shannon, K. (1998). Precision Agriculture: Yield Monitors. University Extension. University of Missouri. 4p.

Corwin DL, Lesch SM (2003) Application of soil electrical conductivity to precision agriculture: theory, principles, and guidelines. Agron J 95:471

Corwin DL, Lesch SM (2005a) Characterizing soil spatial variability with apparent soil electrical conductivity: I. Survey protocols. Comput Electron Agric 46:103–133

Corwin DL, Lesch SM (2005b) Characterizing soil spatial variability with apparent soil electrical conductivity Part II. Case study. Comput Electron Agric 46:135–152

Corwin DL, Lesch SM, Farahani HJ (2008) Theoretical insight on the measurement of soil electrical conductivity. In: Handbook of agricultural geophysics. [S.l.: s.n.t.], pp 59–83

Dammer KH, Hamdorf A, Ustyuzhanin A, Schirrmann M, Leithold P, Leithold H, Volk TE, Tackenberg M (2015) Target-orientated and precise, real-time fungicide application in cereals. Landtechnik 70:31–42

Ehlert DE, Dammer KH (2006) Widescale testing of the crop-meter for site-specific farming. Precis Agric 7:101–115

Ferguson RB, Hergert GW (2009) Soil sampling for precision agriculture. Extension, EC, 154 p

Grisso RD, Alley MM, Mcclellan P (2009) Precision farming tools – yield monitor, Publication number 442–502. [S.l.]: Virginia Cooperative Extension

Lee WS, Alchanatis V, Yang C, Hirafuji M, Moshou D, LI, C. (2010) Sensing technologies for precision specialty crop production. Comput Electron Agric 74:2–33

Miu P (2015) Combine harvesters: theory, modeling, and design. [S.l.]: CRC Press

Moreira MA (2011) Fundamentos do sensoriamento remoto e metodologias de aplicação, 4th edn. Editora UFV, Viçosa

Novo EML (2010) Sensoriamento remoto: princípios e aplicações, 4th edn. São Paulo, Edgard Blucher, 387 p

Queiroz DM, Sousa EDTS, Lee WS, And Schueller JK (2020) Development and testing of a low-cost portable apparent soil electrical conductivity sensor using a beaglebone black. Appl Eng Agric 36(3):341–355

Rouse JW, Haas RH, Schell JA, Deering WD (1973) Monitoring vegetation systems in the great plains with ERTS. In: Proceedings of the Third Earth Resources Technology Satellite—1 Symposium; NASA SP–351. pp. 309–317

Solie JB, Monroe AD, Raun WR, Stone ML (2012) Generalized algorithm for variable-rate nitrogen application in cereal grains. Agron J 104:378–387

Trimble Navigation Limited (2013) Manage your nitrogen and maximize your profit. Available at: https://www.agrioptics.co.nz/wp-content/uploads/2013/04/rt200.pdf. Access 9 Mar 2020

Varvel GE, Wilhelm WW, Shanahan JF, Schepers JS (2007) An algorithm for corn nitrogen recommendations using a chlorophyll meter based sufficiency index. Agron J 99:701–706

Chapter 9
Control and Automation Systems in Agricultural Machinery

Daniel Marçal de Queiroz, Domingos Sárvio M. Valente, and Andre Luiz de Freitas Coelho

9.1 Introduction

Agricultural machinery has been undergoing enormous progress since the 1990s. Until then, agricultural machinery was largely triggered by mechanical systems. Depending on the function performed by the machine, hydraulic actuators were sometimes also used due to the flexibility of these systems. From about 1990, electronic systems embedded in these devices began to be more commonly used. This reached the entire machinery sector, involving tractors, harvesters, and other agricultural machines and implements. This implementation of embedded electronic systems, such as the Global Navigation Satellite Systems (GNSS), allowed for localized crop management, known as precision agriculture.

Precision agriculture has brought the need for the development of variable-rate application machines, in which the rate of the input to be applied is automatically modified. Yield monitors were created so that harvesting machines, in addition to harvesting the product, could collect data to generate the crop yield map, information that is of great value for decision-making in the following harvests. All these technologies that have been and are being important in the development of precision agriculture are also important in digital agriculture.

This chapter discusses the main characteristics of agricultural machines used in precision agriculture and digital agriculture. Basic characteristics of variable-rate application machines, as well as the issue of robotization of agricultural operations, are presented.

D. Marçal de Queiroz (✉) · D. S. M. Valente · A. L. d. F. Coelho
Department of Agricultural Engineering, Universidade Federal de Viçosa, Viçosa, Brazil
e-mail: queiroz@ufv.br; valente@ufv.br; andre.coelho@ufv.br

© The Author(s), under exclusive license to Springer Nature Switzerland AG 2022
D. Marçal de Queiroz et al. (eds.), *Digital Agriculture*,
https://doi.org/10.1007/978-3-031-14533-9_9

9.2   Variable-Rate Application Systems

Variable-rate application is carried out using a system that automatically varies the rate of the input that is being applied. For this, it is important to know the rate that should be applied at each point of the field. The variable-rate application technology has evolved a lot since the early 1990s. There are machines capable of performing variable-rate application of most different inputs used in agriculture. Two types of variable-rate application systems are generally used: map-based systems and real-time sensor-based systems.

In map-based variable-rate application systems, a map with the prescription of the input to be applied is stored in the machine control system. From the data contained in this map, the current position of the machine and the displacement speed, the system controller sends signals to the actuator that controls the metering of the input. The rate of the input is automatically adjusted as the machine moves in the field to meet the established on the prescription map.

Therefore, in map-based application systems, a prior analysis is required to define the rate of the input that will be applied at each point. This analysis can be performed using the entire database that already exists in the available geographic information system and/or in a survey of georeferenced data specific for the application to be performed. Map-based systems use GNSS receivers to determine the location of the machine, so that its control system seeks in the application map the rate to be applied. Then, the controller sends a signal to the actuator, which adjusts the metering mechanism of the machine. A scheme of the components of a map-based variable-rate application system is shown in Fig. 9.1.

In real-time sensor-based variable-rate application systems, there is no previously defined prescription map. Sensors are used to monitor the soil-plant-atmosphere system as the machine moves in the field applying the input. The controller receives the signal from the different sensors and, based on a decision algorithm, defines the signal that is sent to the actuator. It then adjusts the regulation of the metering mechanism to apply the required amount of input for that moment. A scheme of the components of a real-time sensor-based variable-rate application system is shown in Fig. 9.2.

There is also a third variable-rate application system, in which there is no automatic rate variation system, and the applied rate is manually regulated. The area can be previously divided into management zones. These zones can be physically demarcated in the fields, allowing the operator to identify when to manually change the machine's settings. This system is laborious, reduces operational capacity, is subject to human error, and cannot be adopted when the goal is to vary the input rate point by point. Due to these restrictions, commercial devices use systems that have capacity to vary the rate automatically.

In variable-rate application systems, there is always more than one variable that needs to be measured, and sensors are used for this. Generally, the system controls the flow of the input to be applied and not the rate directly. Therefore, it is necessary to convert the rate to a flow value, which requires that the system has a sensor to measure the machine displacement speed, as shown in Eq. 9.1. To determine the

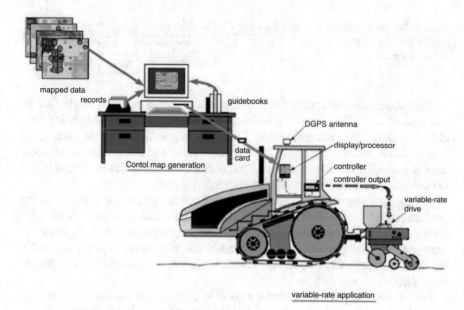

Fig. 9.1 Map-based variable-rate application system. (Source: Ess et al. 2001)

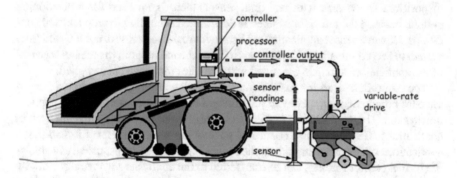

Fig. 9.2 Real-time sensor-based variable-rate application system. (Source: Ess et al. 2001)

operator's travel speed, the same principles used to measure the travel speed of yield monitors can be used.

$$q_i = RsW_s \tag{9.1}$$

where:

q_i = flow of input to be applied by the metering mechanism, in kg/s or L/s;
R = rate of the input to be applied, in kg/ha or L/ha;
s = system's displacement speed, m/s;
W_s = working width covered by the metering mechanism, m.

Fig. 9.3 Feedback control system. (Source: Authors of this chapter)

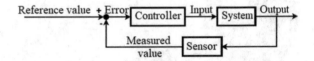

Most control systems work by feedback, whose layout is shown in Fig. 9.3. In this type of controller, there is a value to be reached for the variable being controlled, usually called reference value. The sensor measures the current value of the variable. The difference between the reference value and the current value is the error produced by the control system. The control system modifies the output signal to the actuator in order to decrease the error to a value close to zero.

In real-time sensor-based variable-rate application systems, in addition to the flow and speed sensor, sensors are used to measure variables related to the soil-plant-atmosphere system. The reference value is determined based on the values of these sensors. Therefore, the control system needs to have the capacity to receive the signal or signals from the sensors and determine the rate of the input that should be applied.

In map-based application systems, a positioning system is required to determine the position of the machine in the field. For this purpose, a GNSS system with differential correction is used in order to obtain higher accuracy in the determination of position. In systems with real-time sensor, there is no need for a positioning system, because the rate to be applied is determined according to crop and soil conditions. However, most application systems with real-time sensors use a positioning system to record the applied rates of each input at each point in the area. At the end of the application, a map is generated with the rates that have been applied.

The controllers used for variable-rate application are usually integrated into an onboard computer, equipped with touchscreen to allow the user to set the application system. These controllers enable the connection of the sensors that are used in applications. They can have a connected positioning system antenna for map-based applications or for recording the applied rates. Application maps, usually in shape-file format (vector format), can be transferred to the controller via memory card, by USB device or even be transmitted using a connectivity system if it is available in the field. Controllers are usually able to command multiple actuators.

The actuators receive the signal from the controller and vary the regulation of the input metering. There are different types of actuators, and the choice of type is made according to the input to be applied and the metering/distribution mechanism adopted. Actuators can be linear or rotary and are usually electrically or hydraulically actuated. The most common type of linear actuator is a hydraulic cylinder used to open or close a gate that regulates the passage of the product. Rotary actuators can be electric or hydraulic motors, which actuate an axle, or a pulley, which is attached to the product meter. In the metering of liquid inputs, such as fertilizers or pesticides, the actuators used are the solenoid valves (electric valves), in which the flow rate is often modified by pulse width modulation (PWM). This PWM technique consists of varying the fraction of time that the valve remains closed and the fraction of time that the valve remains open. Commercial devices use pulsation frequency that can vary between 10 and 500 Hz.

9.2.1 Steps for Implementing the Map-Based Variable-Rate Application

The first step in implementing a map-based variable-rate application system is to define whether applications will be varied based on grid or management zones. If it is based on a grid, a regular grid of points must be created. The grid size must be consistent with the working width of the machines that will be used; for example, grids of 5 m by 5 m or 10 m by 10 m can be generated. If a definition of management zones is adopted, these should be defined using information such as yield maps, soil apparent electrical conductivity maps, images obtained by remote sensing, and digital elevation model, among others. Computer programs based on cluster analysis can be used to define these management zones.

Once the grid of points or management zones is defined, the next step is to determine the rates to be applied and generate a prescription map. This map should be generated in the geographic information system, and the file containing that map should be created consistently with the controller of the variable-rate application system.

The generated map should be transferred to the application machine controller. The controller will need to be properly set for the type of input and the type of machine to be used in the application. Once set, the system must be calibrated, since different sensors and actuators make up a variable-rate application system. In addition, the characteristics of the input alter the behavior of the metering mechanisms. Only after calibration, the variable-rate application system will be ready to perform the operation. During the application of the input, it is important that the system stores the application data so that an actual map of that application can be generated. The actual map can be used for traceability of the agricultural product produced and also for the evaluation of the management techniques adopted.

9.2.2 Errors in Variable-Rate Application

Variable-rate application systems may apply different rates from those specified in the prescription map or from the one predicted by the algorithm implemented in the real-time sensor-based application system. This constitutes an error of the application system.

There are several causes of metering errors in variable-rate application systems. For example, if the system is not properly calibrated prior to application, errors may occur due to the change in the behavior of sensors and actuators resulting from environmental variations and changes in the characteristics of the product to be applied. All of these factors affect the meters.

Another factor that can cause errors in the applied rate is to operate the machine outside the range for which it was calibrated. For example, if the calibration was performed for the machine to work at 5 km/h and the operator uses the system in the field at 7 km/h, the speed may change the response between the actuator and the meter. Therefore, it is important that calibration is performed predicting the conditions under which the machine will be used in the field.

A third source of errors is system's response time. The response time is the time interval that it takes from the moment that sensor measures the value of a variable until the meter is applying a rate compatible with the read value. This type of error is associated with the types of sensors and actuators used and the control strategy adopted. The shorter the response time, the lower the error in the system.

The resolutions of the point grid or of the delimited management zones are also sources of errors, as variations may be occurring and the application system may not be responding adequately to these variations.

The complexity of typical systems makes it difficult to determine the results of the interactions of the different errors and whether they will combine the individual errors into an overall large error. If the errors interact badly, the applied rate may be very different from the prescribed rate. Monte Carlo simulation and other techniques can sometimes be used to predict performance, but it is often difficult to make accurate predictions of the errors of such complex interacting nonlinear systems. It is therefore often best to try to minimize each error source in any automation system.

9.2.3   Automatic Section Control System for Input Application Machines

One of the accessories of variable-rate application machines that have had a high level of adoption is the automatic section control system. These systems make it possible to automatically turn off certain sections of the application machine. This prevents the application machine from distributing input where it has already been applied and avoids application where it is not necessary, such as at the end of a plot.

The more irregular the shape of the area and the larger the application width of the machine, the greater the benefit produced by the automatic section control system. Another factor that affects system performance is the number of sections that the system adopts. The number of sections depends on the characteristics of the controller used, and systems with at least five sections are generally adopted. The larger the number of sections, the smaller the area that will receive double the rate and the smaller the waste.

Figure 9.4 illustrates the effect of using automatic section control systems in seeders. Figure 9.4a shows that, in systems without automatic section control, areas are left without seeding or are treated in duplicate. Conversely, Fig. 9.4b shows that, in systems with automatic section control, the areas with double seeding are reduced to a minimum. If the system makes it possible to control each row individually, maximum precision is achieved in the planting system.

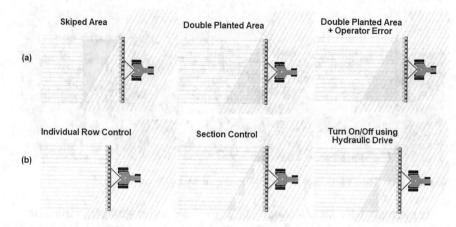

Fig. 9.4 Effect of the use of automatic section control systems in seeders (**a**) without section control and (**b**) with section control. (Source: Runge et al. 2014)

In systems with automatic section control of the area, it is necessary to load a file containing the contour of the area, or the operator must go around the area with the machine before application, so that the system stores the contour coordinates. Section control systems need to be calibrated. Calibration consists of adjusting the time that the system should check ahead to set the time that each section will be turned off. If this time is not properly adjusted, the product flow to a given section will be closed before or after the optimal time, resulting in areas that received no application or those where no input should be applied.

9.3   Steering Assist Systems for Agricultural Machinery

Path-tracking control systems are designed to facilitate the use of machines in the field. Initially, the light bar systems were launched, which indicated to the machine operator whether the machine was swerving to the right or left of the line to be followed. Thus, the operator could correct the direction of the machine. The light bar systems are connected to the positioning system with differential correction, to determine the path taken by the machine. After the launch of these systems, systems with autopilot were developed and commercially marketed.

Autopilot systems allow the machine path in the field to be set automatically. They also allow the path used in a given operation to be recorded, so that subsequent operations are performed following the same path between rows. In addition, it is possible to develop controlled traffic strategies, restricting the areas of machine movement and the consequent restriction of areas subject to soil compaction (Antille et al. 2018).

Fig. 9.5 John Deere's AutoTrac Vision System. (Source: John Deere)

In machines with autopilot, the operator performs maneuvering operations, and from one end of the field to the other, the autopilot system controls the steering system of the set, freeing the operator to other functions. The use of autopilot systems reduces the probability of areas with double application or that do not receive the input, since there is better control of the path to be taken by the machine. In addition, these systems make it possible to work with the same precision both during the day and during the night. Another contribution of these systems is to allow the use of machines with greater working width. Larger machines generally have lower cost per unit of worked area.

Autopilot systems make it possible to work at optimal speed all the time, which means better quality in operation and usually leads to fuel saving, since the engine can work under a better efficiency condition and the operational capacity is usually higher. The use of these systems reduces operator fatigue because they do not need to be so focused on the machine's steering system.

For example, John Deere has a digital vision system that helps the operator keep the operation aligned with the planting rows when spraying, performing cultural practices, or harvesting at higher speeds. This system, called AutoTrac Vision, uses a high-resolution front camera to automatically guide machines through planting rows (Fig. 9.5) and provide their real-time positioning without pre-programmed routes.

9.4 Agricultural Machinery Operation Monitoring Systems

Digital agriculture is supported by three pillars: connectivity between machinery and rural enterprise; the Internet of Things, which allows the collection of data and the transfer of these data to servers in the clouds without human intervention; and sensors and control and automation systems, which equip agricultural machinery.

Companies in the agricultural sector are developing platforms that collect data from various sensors that are made available in the Controller Area Network (CAN) of the machine, such as engine rotation, fuel consumption, travel speed, and position given by a GNSS system, among others. These data can then be transferred to a server in the cloud that processes and turns them into information. The information generated is accessed by users through applications made available for micro-computers, tablets, or smartphones. Examples of these types of platforms are FieldView from Climate Corporation, Operations Center from John Deere, FarmCommand from Farmers Edge, and SGPA from Solinftec.

These platforms provide information on the operation of agricultural machinery. Those responsible for managing operations verify whether what was previously planned is running or if any operational failure is occurring in the field. Platforms also produce maps of applied inputs and yield maps. Also, they facilitate the pre-scription of inputs to be applied and generate the application maps. Automatic weather stations can be installed and connected to the system. The information collected from these stations is taken into account to define the beginning of planting or to verify whether the weather conditions are suitable for the application of agri-cultural pesticides. This improves the efficiency of operations.

To transfer the data, the digital agriculture platforms make use of drives, which extract information of the machine through the CAN network and transfer it to a tablet. The data stored on this device are transferred in real time to a centralized system or directly to servers in the cloud if there is connectivity in the field. If this connectivity does not exist, the data is stored on tablets for later transfer. Depending on the system, instead of using the drive, the onboard computer is used, which con-nects to the machine's CAN network and collects the information. These are sent, via telemetry, to a center, which uploads the data in servers in the cloud.

9.5 Robotics in Agriculture

Several factors are leading to the robotization of agriculture. First, the lack of labor in the field, as the field work is hard and workers prefer to work in cities. In addition, there is an aging of the rural workforce in various parts of the world, causing an increase in the average age of rural workers. Machines with high degree of automa-tion are easier to operate. The robotization of agriculture is now a necessity if the desire is to maintain the food security of the world's population. Figure 9.6 shows two models of autonomous tractors and robotic platforms presented by the industry for agricultural applications.

Another benefit of robots is the gain of quality in operations. Generally, robotic systems perform the activities with greater perfection, resulting in gains of quality and yield. Examples of the use of robots are those being used for operations in experimental fields for genetic improvement of agricultural crops and those used for work in protected environments. Robots equipped with cameras and sensors can perform measurements of a larger number of variables, automatically and with less subjectivity. In addition, with the use of robotic machines, it is possible to apply the inputs in the quantity needed and in the place where they are really necessary, lead-ing to the practice of more sustainable agriculture. Technological development in

Fig. 9.6 Autonomous tractors and robotic platforms for agricultural crops. (Source: The authors)

the areas of sensors, microprocessors, and actuators, together with advances in information technology and communication, has enabled the development of more efficient and better quality machines.

The first autonomous tractors begin to be presented by agricultural machinery manufacturers. Robots for mechanical cultivation operations and application of agricultural pesticides and for harvesting are already being made available in the market. Unmanned aerial vehicles equipped with agrochemical application systems are beginning to be adopted in agriculture. The trend is that in the coming years, many agricultural operations will begin to be fully executed by robots or unmanned aerial vehicles.

Another application of robotics in agriculture is in crop monitoring operations. Robots equipped with cameras and other sensors can collect data with much higher resolution than when using aerial platforms. After being processed, the data can lead to more accurate diagnosis and prescription in the use of inputs.

The use of robots in agriculture will mean an increase in the amount of data collected during agricultural operations. Digital agriculture will benefit from these data, improving the decision-making process. In addition, digital agriculture platforms will be able to transfer input application maps directly to robots, resulting in time and labor savings.

9.6   Digital and Precision Agriculture
for Smallholder Farmers

As shown in this chapter, various machines, devices, and tools for digital and precision agriculture are available to farmers. However, some factors have restricted their adoption by farmers who cultivate small areas or with low potential for investments in technology. These factors include the large dimensions, the high cost of acquisition, and the need for trained professionals to operate and maintain these instruments. Therefore, it is common to associate precision agriculture and digital agriculture with large producers. However, it is important to note that smallholder farmers are responsible for feeding about a third of the world's population.

The recent popularization of the concepts of components, platforms, and open-source hardware and software programs has enabled the development of equipment with affordable prices for small producers. Examples of this are affordable sensors, controllers, and micro-controllers such as single-board computers and GNSS modules. In the laboratory and field testing phase, several prototypes of equipment have been reported in the literature: for instance, the autopilot based on low-cost GNSS receiver (Alonso-Garcia et al. 2010), the acquisition system for plant canopy monitoring (Fisher and Huang 2017), the sensor for measuring soil apparent electrical conductivity (Queiroz et al. 2017), the computer program for embedded system with the function of data spatial variability analysis (Coelho et al. 2018), and the data acquisition system for energy analysis in agricultural tractors (Santos et al. 2018).

The authors of this chapter have been working on the development of a map-based controller for variable-rate seeding and fertilizer distribution using the following tools: a Beaglebone Black single-board computer running a computer program developed in Python language that controls all operations necessary for variable-rate application, a low-cost GNSS module to obtain the machine's position in the area, and manual seeder with Knapik wheel (Knapik Indústria Mecânica, Porto União, Brazil), modified to receive the controller. Figure 9.7 shows a prototype during a field test.

In addition to performing and controlling the distribution of seeds and fertilizers, the controller contains a Global Service Mobile (GSM) module that makes it possible to send operation data, using the internet service provided by mobile operators. The data are stored on a server, and the information is presented to the farmer in the form of maps (Fig. 9.8). The cost of acquisition of the components needed for developing the controller was approximately US$430.00.

Abbreviations/Definitions

- Variable-rate application machines: Machines capable of automatically varying the dosage of agricultural inputs.
- Map-based variable-rate application: A map with the prescription of the input to be applied is stored in the machine control system. The machine automatically

Fig. 9.7 Seeder-fertilizer with low-cost controller developed for variable-rate seeding and fertilizer distribution. (Source: Authors of this chapter)

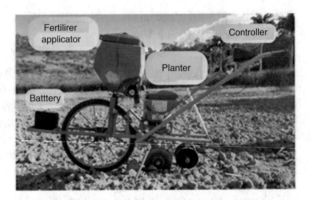

Fig. 9.8 Map of displacement speed of the seeder-fertilizer with the controller, after a seeding operation. (Source: Authors of this chapter)

changes the application rate based on the map and on the position the position obtained by a GNSS sensor.

- Real-time sensor-based variable-rate application: There is no previously defined prescription map. Sensors are used to monitor the soil-plant-atmosphere system, and microprocessors are used to calculate the input to be applied as the machine moves in the field doing the application.
- Yield monitor: System used in harvesters that is capable of collecting data to generate crop yield maps.
- Pulse Width Modulation (PWM): It is a signal where you control the fraction of time that signal is kept high and low. In digital agriculture, PWM is used to control the flow of the product being applied by controlling the percentage of time the valve should be on and off.

- Section control: This system makes it possible to automatically turn off certain sections of the application machine. This prevents the application machine from applying input where it has already been applied and avoids application where it is not necessary, such as at the end of a plot.
- Steering assist system: Control system designed to facilitate the machine orientation in the field.
- Light bar: System connected to the positioning system with differential correction to determine the path that should be taken by the machine. In this system, the machine operation is manual.
- Autopilot: Systems to automatically drive the machine over the field.
- Internet of Things: Sensors connected in things and connected to internet to transfer data in real time. It allows to collect machine data and transfer it to a cloud server without human intervention.
- Controller Area Network (CAN): It's a communication standard that makes it possible for microcontrollers and devices to communicate one to each other without the need for a host computer.
- Global Service Mobile (GSM): Module that makes it possible to send operation data, using the internet service provided by mobile phone operators.

Take Home Message/Key Points

- Variable-rate application is carried out using a system that automatically varies the rate of the input that is being applied.
- Two types of variable-rate application systems are generally used: Map-based systems and real-time sensor-based systems.
- Digital agriculture is supported by three components: Connectivity between machinery and rural enterprise; the Internet of things, which allows the collection of data and the transfer of these data to servers in the clouds without human intervention; and agricultural machinery sensors, control, and automation systems.
- The use of robots in agriculture will increase the amount of data collected during agricultural operations. Digital agriculture will benefit from these data, improving the decision-making process.

References

Alonso-Garcia S, Gomez-Gil J, Arribas JI (2010) Evaluation of the use of low-cost GPS receivers in the autonomous guidance of agricultural tractors. Span J Agric Res 9:377–388

Antille DL, Chamen T, Tullberg JN, Isbister T, Jensen TA, Chen G, Baillie CP, Schueller JK (2018) Controlled traffic farming in precision agriculture. In: Precision agriculture for sustainability, Burleigh Dodds Series in Agricultural Science. Burleigh Dodds Science Publishing, Cambridge, pp 239–270

Coelho ALF, Queiroz DM, Valente DSM, Pinto FAC (2018) An open-source spatial analysis system for embedded systems. Comput Electron Agric 154:289–295

Ess DR, Morgan MT, Parson SD (2001) Implementing site-specific management: map-versus sensor-based variable rate application. Pub. n° SSM-2-W, 2001. Disponível no Site-Specific Management Center, Purdue University, West Lafayette

Fisher DK, Huang Y (2017) Mobile open-source plant-canopy monitoring system. Mod Instrum 6:1–13

Queiroz DM, Lee WS, Schueller JK, Santos EDT (2017) Development and test of a low cost portable soil apparent electrical conductivity sensor using a Beaglebone Black. In: SPOKANE, St. Joseph, MI., July 16–July 19, 2017, Washington. International Meeting of Agricultural and Biological. American Society of Agricultural and Biological Engineers, St. Joseph

Runge M, Fulton J, Griffin T, Virk S, Brooke A (2014) Automatic section control technology for row crop planters. Extention of Alabama A & M and Auburn Universities. Bulletin ANR-2217

Santos DDC, Mantovani EC, Barbosa AM, Mewes WLDC (2018) Uso de linguagem C++ em computador embarcado de alta capacidade para eficientização energética em tratores. Revista Brasileira de Milho e Sorgo 17:340–352

Chapter 10
Digital Irrigation

Roberto Filgueiras, Lucas Borges Ferreira, and Fernando França da Cunha

10.1 Introduction

The constant search for higher yields and lower production costs is an objective present in twenty-first-century agriculture. Installation of irrigation systems in areas with low levels and/or irregular distribution of rainfall is a major factor to increase yield. In these cases, irrigation can provide both quantitative and qualitative gains, besides conferring greater guarantee of production. However, irrigated agriculture requires a large volume of water, which is why the installation of irrigation systems must be accompanied by an adequate management program. Moreover, correct management of irrigation, besides promoting water savings, can promote yield gains, decreased risk of diseases, higher fertilization efficiency, and energy savings. Therefore, the factors that guide the decision to irrigate must be monitored in detail.

Irrigation management, conventionally carried out in rural properties, is based on climate. In this type of management, water demand is obtained for the entire agricultural field based on reference evapotranspiration (ETo) and crop coefficient (Kc), in addition to other adjustment coefficients, such as the water stress coefficient (Ks). These data are used to generate an average value of water depth to be applied in the entire field. However, the factors that guide the decision to irrigate, that is, soil, plant, and climate conditions, show spatial and temporal variability. Therefore, a more detailed monitoring of these factors is needed, aiming to meet not only an average water demand per field but also a specific demand for each site. Based on these principles, on the growing need for greater profitability in agriculture and on the fact that water is a finite resource, agricultural producers are resorting to technology to irrigate with greater assertiveness.

R. Filgueiras (✉) · L. B. Ferreira · F. F. da Cunha
Department of Agricultural Engineering, Campus of the Federal University of Viçosa, Viçosa, Minas Gerais, Brazil
e-mail: roberto.f.filgueiras@ufv.br; lucas.b.ferreira@ufv.br; fernando.cunha@ufv.br

In this context, together with precision agriculture and digital agriculture, the terms precision irrigation and digital irrigation are born, which encompass the technologies that allow the crop to be monitored and managed with greater precision, avoiding waste, increasing profitability, and making the production process more sustainable. Precision irrigation brings from precision agriculture the idea of treating differently what is different, giving spatial detailing to the actual water demand in the agricultural field and promoting the zoning of regions that need more and less water. One of the advances in this area occurred with the increasing use of aerial and orbital images in agriculture and, concomitantly, with the beginning of estimates of actual evapotranspiration with satellite images, based on both empirical and physical algorithms. However, as precision agriculture, precision irrigation is focused not only on the differentiated prescription of crop irrigation requirements but also on the application of the water depth according to the demand of each site, using variable-rate application equipment.

Digital irrigation can be defined as a segment of digital agriculture that focuses on the management and monitoring of irrigated agriculture, consisting of several technologies that assist the rural producer in its irrigated systems. Like digital agriculture, digital irrigation uses software and devices that collect, store, and analyze data from different sources to make the system automated and provide information for strategic decision-making.

Although precision irrigation and digital irrigation refer to different concepts, as already mentioned, they are used together many times, especially in large rural properties that, in addition to requiring the automation of irrigation systems to facilitate their operation, need a differentiated control of the system in space. Moreover, in practical situations, the term digital irrigation is commonly used considering also the principles of precision irrigation. As both segments are terminal technologies in the production process and focus on the efficiency of irrigated systems, in this chapter, when digital irrigation is mentioned, the reader must understand that this term may be encompassing aspects related to precision irrigation.

One of the characteristics of irrigation digitization is to bring convenience and safety to the rural entrepreneur by collecting and interpreting data regarding the water need of plants in the field. For this, the system can use sensors in the field, aerial images, satellite images, equipment for processing and data analysis, as well as other technologies that will allow farmers to accompany the irrigation data of their farm, in both time and space. In addition, the use of data science and machine learning techniques plays a prominent role in the process of transforming data into useful information for system management.

10.2 Irrigation Project and Systems

In general, a good irrigation system should be able to apply water at the appropriate time and quantity to meet the water need of the crop. Given the temporal and spatial variability of the water demand of crops, the system should be designed in such a way that it is possible to consider such variability at the time of irrigation. The variation in irrigation depth over time can be easily controlled by irrigation time or by the displacement speed of the irrigation equipment, in the case of traveling equipment. However, for the spatial variability of the need for irrigation to be considered, some aspects, which are discussed in the sequence, must be taken into account in the creation of the project.

Regarding various irrigation systems available, they have different capacities to apply water considering the spatial variability of water demand. However, even in traditional irrigation systems, such as conventional sprinkler, it is possible to design the project aiming at the highest precision of water application.

10.2.1 Data Acquisition

Prior to the creation of the irrigation project, it is necessary to collect information regarding soil, climate, and crop. As for soil, it is necessary to know its physical-hydric properties, such as field capacity (FC), permanent wilting point (PWP), and soil bulk density (Ds). These variables are used to calculate the soil water storage capacity, necessary to define the maximum irrigation interval, besides being used in the irrigation management stage, since this capacity will influence the water consumption of the crop. Knowledge on the local climate and crop is necessary in order to estimate the water need of the plant in the period of maximum demand, which should be met by the irrigation system. For this, one must know the maximum evapotranspiration of the crop of interest (ETc) for the site and period in which the cultivation will be carried out. Maximum ETc is obtained based on the maximum reference evapotranspiration (ETo) and the maximum crop coefficient (Kc).

The abovementioned soil data must be obtained in such a way as to represent the different types of soil present in the field. Ideally, FC, PWP, and Ds data could be sampled in a grid, interpolating these variables later. From these data, it is possible to define management zones, that is, areas with homogeneous characteristics. After this definition, the project should be designed with a view to allocating the different management zones in irrigation units with independent operation (sectors), which will allow the application of different irrigation depths in each zone. However, in practical situations, the collection of large amounts of samples to obtain FC, PWP, and Ds is a limitation for the adoption of this method. Thus, another possibility would be the definition of management zones from variables easily collected, a common practice in precision agriculture (Gavioli et al. 2019). For this, it is possible to use data of soil texture and electrical conductivity, among others. From these management zones, it is possible to perform a directed sampling of FC, PWP, and Ds, which reduces the number of samples, since the management zone is considered homogeneous. However, if this approach is not possible, the producer can collect samples to determine FC, PWP, and Ds in a more general way but seeking to separate areas with soils of noticeably different textures.

For the calculation of maximum ETc, obtaining the maximum ETo (maximum ETo value associated with a given probability of occurrence) constitutes the most difficult step. On the other hand, Kc values can be easily found in the technical literature. The maximum ETo is usually determined by means of historical ETo series obtained from data measured at a meteorological station near the site where the irrigation system is installed. However, there is not always a weather station near the site of interest. Moreover, obtaining the maximum ETo from meteorological data is usually a laborious and time-consuming activity.

In Brazil, in order to provide easier access to maximum ETo data for irrigation project, Dias (2018) obtained, from data of orbital remote sensing and other gridded products, spatialized values of maximum ETo for the entire Brazilian territory (Fig. 10.1). Maximum ETo values are available for each month of the year and on an annual scale. Such information is available for free (www.gesai.ufv.br) and can be used quickly and without the need for data collection at weather stations. Similar information can be achieved for any other place in the world using the methodology employed by Dias (2018), who used data from meteorological stations and MODIS sensor to model maximum ETo.

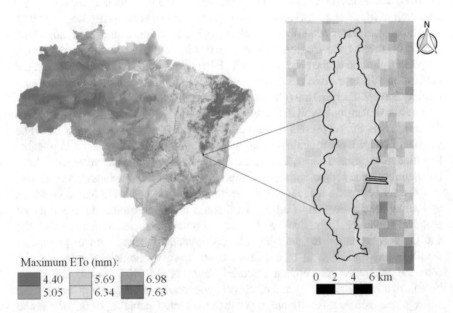

Maximum ETo (mm):

4.40	5.69	6.98
5.05	6.34	7.63

0 2 4 6 km

Fig. 10.1 Maximum annual ETo for irrigation project purposes in Brazil, highlighting the irrigation perimeter of Gorutuba (Nova Porteirinha-MG). (Source: Authors of this chapter)

10.2.2   *Irrigation and Automation Systems*

In digital irrigation, the advances of irrigation systems and the technologies embedded in them play a prominent role, since they provide a higher level of automation and, in some cases, allow the variable rate application of water. Among the innovations currently available in the market, some, which we believe to be in the main ones, are presented in subsequent topics.

10.2.2.1   Center Pivots with Variable-Rate Irrigation

Center pivots are traditionally designed to apply the same irrigation depth throughout their coverage area. Thus, it becomes difficult to vary the irrigation depth along the area. However, it is still possible to do some basic controls on a traditional center pivot, for example, apply different depths on the two halves of a pivot with a different crop in each half.

Due to the need to increase irrigation accuracy, pivots with variable-rate application have been developed, known as variable-rate irrigation (VRI). These devices make it possible to apply water in different quantities within the irrigated area. There are two types of VRI: one based only on the control of the pivot's displacement speed along its radial trajectory, increasing or reducing the applied depth, and another that, in addition to being able to vary the pivot's displacement speed, can turn each sprinkler or sprinkler group on and off independently. The first type of VRI is usually called speed control VRI or sectorial VRI, and the second type is called zone control VRI.

Zone control VRI has greater flexibility to vary the irrigation depth in the irrigated area and can divide the area into a large number of independent irrigation zones. In the case of the zone control VRI with individual control of each sprinkler, the detailing is even greater than that achieved when controlling groups of sprinklers. On the other hand, the speed control VRI allows only the variation of irrigation depths in pizza slice-shaped areas. The water distribution patterns of the two types of VRI are illustrated in Fig. 10.2.

10.2.2.2   Corner Pivot

The center pivot irrigates a circular area, thus leaving a border area, which in most cases is cultivated in a rain-fed system. Yield in these areas tends to be lower than that of irrigated areas, since they depend solely on rainwater. To solve this problem, a solution developed was the articulated arm system, called corner.

Emitters in the corner are in greater number than those present in conventional spans and composed of hydraulic valves that operate open and closed, according to the water need of the target to be irrigated. If the corner is open (extended), all emitters are irrigating. However, as the corner is closed (retracted), the hydraulic valves

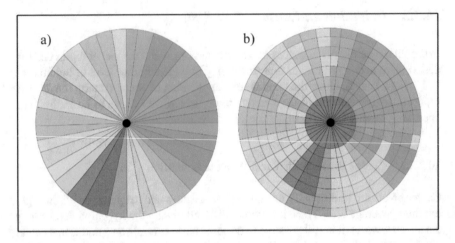

Fig. 10.2 Types of VRI center pivot: (**a**) speed control and (**b**) zone. (Source: Authors of this chapter)

Fig. 10.3 Center pivot with the corner system: (**a**) Area increment in relation to a conventional center pivot and (**b**) corner connection point to the center pivot. (Source: Authors of this chapter)

are also closed and interrupt the application of water. The entire embedded technology system is programmed and controlled by means of a GNSS RTK with centimeter accuracy, which promotes perfect synchrony between the corner extension and the valves.

Figure 10.3a shows the increase of area obtained with the corner in a conventional center pivot, from 85 to 102 ha, which implies an increase of around 20% in the cultivation area. Figure 10.3b shows the detail of the corner connection to the last tower of the center pivot (Oliveira et al. 2018).

Center pivots with the corner system must be well designed, since at some point during the irrigation, the corner span, which was previously attached to the center pivot spans, opens, substantially increasing the number of emitters in operation.

There are options of corner on the market that reach up to 81 m in length. The increase in the number of emitters in operation requires a higher flow rate, increasing the speed of water in the pipes and, consequently, the head loss. With the increase in head loss, greater power of the motor pump is required to increase the pumping head of the system as a whole. If the center pivot with corner is not well designed, a pressure drop may occur in the entire system when the corner begins to operate, so the crop is not properly irrigated.

10.2.2.3 Automated Fertigation

The preparation of the fertilizer solution and the quantitative and qualitative control of its injection into the irrigation system are costly tasks. Thus, this combination of services can be automated, improving the operational performance of fertilization. Fertigation can also maximize yield and reduce fertilizer costs by providing plants with nutrients in the required amounts at the right time.

There are automatic devices on the market to control fertilizer injection in the irrigation line. These fertigation systems are composed of metering pumps for controlling the injection of fertilizers. Generally, the fertilizer dose is automatically controlled by the electrical conductivity (EC) of the solution to be applied, but there are other devices that also consider the pH. In addition, these devices can monitor the reservoir level, pump working pressure, dissolved oxygen in the solution, level of the diluted solution, and solution temperature.

EC and pH monitoring is indispensable in fertigation. The lack of control of the level of these water quality parameters can lead to death of crops. Automatic monitoring of these parameters and correction of these values when they exceed optimal levels are needed. Automatically, these systems can control the injection of two types of solutions, injection of acid for pH reduction, and injection of water to reduce EC.

10.2.2.4 Irrigation Monitoring and Remote Control

Currently, there are several commercial solutions, through smartphones and computers, for monitoring and remote control of irrigation systems, especially in the case of center pivots. The level of control varies depending on the commercial solution chosen. In general, the available products are able to turn irrigation systems on and off, in addition to monitoring whether the irrigation device is on and informing the position and direction of its displacement in the field (in the case of pivots). Systems can also detect theft of electrical cables and monitor pumping installations, flow rate, and system pressure. Commercial solutions include BaseStation (Valley), FieldNET (Lindsay), Kube (Krebs), and Gavish Connect (Agrosmart/NaanDanJain).

Monitoring and control systems allow users to interact with irrigation equipment through an internet connection. However, as irrigation systems are often located in areas without internet coverage, automation systems can use radio communication between the equipment present in the irrigated area and the rural company head-quarters, where there must be an internet connection so that the user can control the system from anywhere through a smartphone, a tablet, or a computer. There are also other possibilities for improving the connectivity of the rural company, such as the use of satellite internet or the creation of own internet networks within the company. Connectivity is a subject of importance for digital irrigation, since the information collected in the field needs to be transmitted to a central unit, where the collected data can be analyzed and transformed into useful information for irrigation management.

10.3 Irrigation Management

Knowing how to manage the irrigation of a crop is a crucial point to leverage the profitability of production, since both excess and shortage of water are harmful. Excess irrigation increases the cost of production, due to both the energy needed for pumping and fertilizers that can be lost by leaching, besides possibly resulting in a loss of yield. The loss of yield can occur due to hypoxia and leaching of nutrients essential for plant growth, which can also cause damage to the environment. However, water deficit in plants causes stress, resulting in decreased yield and quality and, consequently, making them more susceptible to pests and diseases. Thus, finding the most efficient form of irrigation management is an important point not only for the profitability of the activity but also for the sustainability of the system, in general. Therefore, the rational use of irrigation is a key point for the maintenance of agricultural activity in the future, since water resources are increasingly scarce. In this context, irrigation management will be introduced in the assumptions of digital irrigation.

The management of digital irrigation is based on the connectivity of sensors and actuators, sending and receiving information in an automated way, based on the concepts of Internet of Things (IoT). This large data transmission makes it possible to act in the monitoring of irrigation remotely and in real time. IoT platforms enable complex and intelligent decision-making, which are supported by data analysis and machine learning algorithms. The large amount of data (big data) received from the field generates new possibilities of understanding the soil-water-plant relationship, generating localized and accurate information for even more efficient irrigation management. Still, it is a challenge to treat all of this collected data, even using cloud computing and intelligent decision-making systems. With the help of information that is generated by digitization of the irrigated agriculture system, the farmer/engineer has other possibilities to understand which factors most influence the crop in the field, answering questions that were previously impossible to solve.

In an integrated decision-making system that considers all variables that influence the response to irrigation, one should not fail to consider the spatial variability of the cultivation area. Because of this factor, sensors must be distributed strategically in the field, allowing data to be acquired for the recognition of the water demand of the crop and, thus, trigger the system differently in space and time, that is, to perform variable rate irrigation.

10.3.1   Possible Approaches to Irrigation Management

Basically, there are three methodologies established to perform irrigation management: plant-based, soil-based, and climate-based. Plant-based management is used almost exclusively in research media mainly because of the high cost of equipment and the high complexity involved in the analysis of variables. However, climate- and soil-based managements are very widespread, each with its particularities. In conventional climate-based management, meteorological data from weather stations are used to estimate ETo and, subsequently, to determine the average net irrigation depth to be applied to the agricultural field. On the other hand, in soil-based management, sensors are used to measure soil moisture at the depth of the crop root zone. However, these conventional methodologies do not take into account spatial differences in an agricultural field. In this context, the spatialized estimates of soil moisture, ETo, and actual evapotranspiration of crops will be presented in the next topics, in addition to presenting the applications of these estimates in digital irrigation.

10.3.2   Soil-Based Management

In soil-based irrigation management, it is determined when and how much water to be applied using data of the water available in soil for plants. Soil moisture can be estimated manually by the gravimetric method, also known as the standard oven method. Unfortunately, it is not feasible to automate the method, an indispensable factor in digital irrigation. This method becomes impractical for this purpose, since frequent measurements are necessary for real-time decision-making. In order to meet the demand for more frequent and cheaper data, several soil moisture sensors have been developed, including sensors capable of transferring data via wireless connection. This makes it feasible to install several sensors in the field with automated collection of measurements, building a data network in order to understand the spatial behavior of water in the soil. Electronic sensors to monitor soil moisture have been widely used, especially when the idea is to collect large amounts of data in time and space, since these sensors can be connected to dataloggers and the data transmitted via wireless. The transmitted data can be accessed on a computer or even on a smartphone, enabling decision-making from the collected data.

When the goal is to carry out irrigation management using soil moisture sensors, the first question to be asked is, "How many sensors should be installed in order to accurately represent the spatial dynamics of water in the root zone of the crop?" This question should be asked because the representativeness of a sensor that measures soil water content or even the tension at which water is retained varies according to the soil, and a prior survey is necessary to estimate the spatial dependence of the measurements of these sensors in the agricultural field. Therefore, the determination of homogeneous management zones is interesting and can help in the installation of these sensors in the field, even if there is not yet a standard number of sensors that must be installed per management zone, which can vary considerably according to soil characteristics. Another important factor that should be taken into account is the location of these sensors in the field, aiming to optimize their position in order to minimize the number of sensors as much as possible and seeking to install them in places that will not hinder cultural practices.

The main idea is that the collection of data from moisture sensors generates spatialized information of soil water content, as illustrated in Fig. 10.4. In the process of converting this current moisture into an irrigation depth, spatialized information on soil bulk density, permanent wilting point, and field capacity are also necessary.

Soil moisture measurements at the irrigation management level are still very dependent on field sensors, since data from orbital remote sensing for this purpose still have low spatial resolution and some limitations that make it difficult to obtain accurate data, not being indicated for irrigation management. Although soil moisture estimates from orbital data, aimed at management of irrigated systems, are still at research level, this approach has great future potential for use in irrigation.

10.3.3 Climate-Based Management

ETo is the water demand exerted by the climate on a reference surface. Therefore, it is an important information for those who want to perform climate-based irrigation management, since it is through this variable that the daily net irrigation depth to be applied in the crop is obtained. However, for the estimation of ETo, considering the method established by the Food and Agriculture Organization (FAO), the so-called

Fig. 10.4 Spatialized (hypothetical) values of soil moisture as a function of the percentage of field capacity (FC) in a center pivot. (Source: Authors of this chapter)

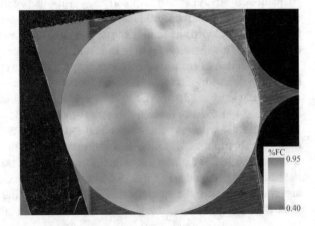

Penman-Monteith method, a weather station is required to capture information on wind speed, solar radiation, temperature, and relative humidity. It is important to consider the spatial variability of the input parameters of the ETo calculation equation, since a single weather station may not represent a given region with reliability.

In this context, ETo products that consider the spatial variability of meteorological elements have been developed, which are created and validated based on data from weather stations of government institutions. These ETo products are created using gridded databases of climatological normal, together with the use of satellite images, which consider the spatial dynamics of clouds and meteorological parameters. The models used to build the ETo products recognize the patterns between input data and weather station data in the most diverse regions of a country and later spatialize ETo based on machine learning models, making the information available to any region. This information can be made available at the level of agricultural fields, which will give greater sensitivity and reliability to the reality of the reference water demand.

It can be said that the consideration of the spatial dynamics of ETo already characterizes a differentiated treatment according to the demand of different agricultural fields. This constitutes an option of climate-based management with greater inclusion in precision irrigation. It is worth pointing out that a farmer—or even a service provider company—that performs climate-based irrigation management using the ETo estimated with a weather station for a large territorial extension and does not consider in its recommendations the variability of ETo in space can accumulate deficit or excess irrigation throughout the season.

10.3.4 Actual Crop Evapotranspiration

The actual evapotranspiration of a crop is the amount of water transferred from the surface to the atmosphere by evaporation and transpiration, under the actual conditions of atmospheric demand, soil moisture, and phenological stage of the crop. Thus, the variability in an agricultural field related to soils, topography, nutrients, and growth rate of the crop, among others, causes the water demand of plants in each locality to be also different. This circumstance makes it important to know the actual evapotranspiration of crops in a spatialized way.

Spatial estimation is performed with the support of ground monitoring satellites and/or unmanned aerial vehicles (UAVs). This is because instruments and methodologies for measuring actual evapotranspiration, such as lysimeters, Bowen ratio-energy balance, and Eddy covariance techniques, are not feasible for the spatial understanding of the agricultural field. These conventional techniques are not directly applied to large areas due mainly to the need for a large network of costly instruments to be able to represent the natural heterogeneity of the surface.

With a view to spatial representation of actual evapotranspiration, researchers have created several algorithms and methodologies to obtain this parameter through satellite images, including the Surface Energy Balance Algorithm for Land

(SEBAL), the Mapping EvapoTranspiration at High Resolution and Internalized Calibration (METRIC), the Simplified Surface Balance Energy (SSEBop), the Two-Source Energy Balance Model (TSEB), and the Simple Algorithm for Evapotranspiration Retrieving (SAFER). For most part, these algorithms consider the equations of the energy balance to estimate evapotranspiration through satellite images. The great advantage of using images is the possibility of estimating evapotranspiration for large areas (Fig. 10.5), systematically over time and at a much lower cost than conventional techniques. In the agricultural context, the estimation of evapotranspiration through satellites has been validated, with data measured in the field, especially on irrigated surfaces.

Recently, with the popularization of UAVs and the progressive increase in research and services with this instrument, there have been studies aiming to estimate evapotranspiration using cameras sensitive to the thermal and/or near-infrared and visible spectrum, such as the study conducted by Santos et al. (2020). These studies allow the actual evapotranspiration of the crops to be estimated in a high level of detail, very high spatial resolution, and a frequency of imaging following the needs of irrigation management. However, the approaches of actual evapotranspiration with orbital images are more accessible and, at least to date, have a greater number of studies and validations in the field.

The modeling of actual evapotranspiration using spectral data, from both satellites and UAVs, using machine learning techniques, is becoming more common in the field of research. There is also a tendency to become common in rural properties, serving in the future as a source of information for digital irrigation.

The estimation of spatially distributed evapotranspiration in the scope of irrigation management will play a fundamental role in the definition of homogeneous evapotranspiration zones, in order to provide information for decision-making, especially in VRI irrigation systems.

10.3.5   Irrigation Management in VRI Center Pivot

The VRI system allows the necessary irrigation depth to be applied at each location, optimizing and allocating water where it is really needed. Variable-rate prescriptions can be integrated into most irrigation systems, but it is still a technology that

Fig. 10.5 Actual evapotranspiration estimated by SEBAL for 33 center pivots on a large agricultural property located in the western state of Bahia, Brazil. (Source: Authors of this chapter)

is under further development and will become more frequent as digital irrigation and precision irrigation become more present in the field.

Applying water at variable rate has numerous agronomic and economic benefits, not being restricted to only the right water depth for each locality. In addition, it can be highlighted that the expenditure on fertilizer may be lower, since water and nutrients are applied only in places of interest, as the system can be used for fertigation. However, management with VRI is more likely to obtain higher results when the area has a large variability of topography and soil. However, it is recommended for any situation, and the VRI system can even be adapted to conventional center pivots.

To start the management with center pivots with VRI, one can define the variability of the area and then define management zones, so that it is possible to define the maps of recommendation for irrigation management. Often management zones are defined based on information of apparent soil electrical conductivity, which is correlated with data that spatially vary within the area, such as texture, water retention capacity, and salinity, among others. However, nothing prevents the use of other information to define the management zones, such as relief information and yield maps.

Several studies have been seeking the best layers of information to use them in the definition of management zones, which leads to a continuous evolution in the methodologies, also because this is an area that advances rapidly as new data acquisition equipment is launched. The prescription of irrigation for VRI system can be stratified by management zones, being defined by moisture sensors distributed throughout the different zones or, also, by satellite images used to estimate the actual evapotranspiration of crops. This latter approach is under development and is more used in research studies.

10.3.6 Monitoring

Crop monitoring plays a fundamental role in the success of the agricultural activity, since it is through this procedure that possible problems in the field will be encountered. The main question of monitoring an area and succeeding in this function is related to the identification of problems before they cause economic damage. Based on this premise, the monitoring of areas in a conventional way, that is, walking along the agricultural field to monitor it, is not such a complex task in relatively small areas; however, in larger areas, it becomes an obstacle to the success of the activity. In this context, agricultural monitoring with the use of images from UAVs and satellites has been an efficient option for both small and large areas.

Satellites have the advantage of systematically acquiring data from large areas, with low-cost or even free options. These devices have been used as a source of agricultural information for some time. The Landsat constellation, which had its first satellite launched in 1972, was the main responsible for this to occur, being still widely used today.

Agricultural monitoring through images has been shown to be relevant for the observation of spatial patterns of vegetation, in which one can identify the most fragile points of the crop, saving time in field inspections. Although it is difficult to identify which biotic or abiotic factor is causing a particular problem in the crop, satellite images point to the sites that deserve the most attention and should be checked in loco, which supports and guides the farmer who uses these data. However, there are some spatial patterns that occur in agricultural crops, mainly with center pivot systems, which make it possible to infer about the cause of the problem by interpreting images. Generally, spatial patterns that allow this identification of problems are associated with the malfunctioning of the irrigation system. In this case, spatial patterns have geometric shapes that are well defined or repeated throughout space, such as those observed in Figs. 10.6 and 10.7. Both figures present the normalized difference vegetation index (NDVI), which allows inferences to be made about the condition of the vegetation. NDVI values range from −1.0 to 1.0, and the closer the index is to 1.0, the greater the vegetative vigor. Understanding the values of this index, and of others, makes it possible to locate regions with plants that are suffering some form of stress in an agricultural field.

It is possible to see in Fig. 10.6 that there is an X-shaped pattern in the center pivot that has lower NDVI values. This can be attributed to the extension of the corner system, since the spatial pattern in this situation is extremely characteristic of its functioning. As previously reported, when the corner starts to open there is pressure drop throughout the system, which does not adequately irrigate the crop.

Two other spatial patterns characteristic of problems in an irrigation system are those presented in Fig. 10.7. Figure 10.7a shows a pattern of strips with greater vegetative vigor that is repeated along the center pivot cultivated with sugarcane. This pattern is not common in images of center pivots, as it points to problems in the operation of the equipment itself. The spatial pattern in Fig. 10.7a indicates that the center pivot is failing to pressurize during operation or is moving its spans before being properly pressurized. In the first case, there may be some problem regarding the supply of electricity because, if there is an oscillation of energy, there may be constant changes in the pumping head, resulting in poor distribution of the applied

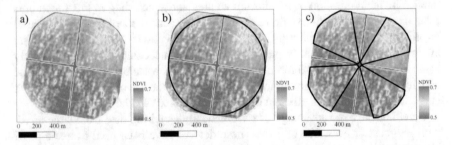

Fig. 10.6 Images with spatial distribution of NDVI: (**a**) area irrigated using a center pivot with the corner system, (**b**) delimitation of the area that would be irrigated by a center pivot without the corner system, and (**c**) X-shaped pattern with lower NDVI values. (Source: Courtesy of AgriSensing)

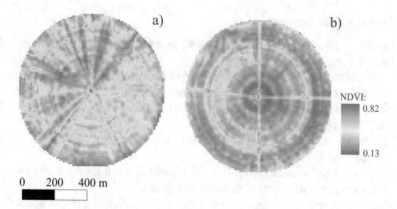

Fig. 10.7 NDVI variation in an area irrigated using a center pivot: (**a**) Image with radial patterns of low and high NDVI values and (**b**) image with circular patterns of low and high NDVI values. (Source: Alface et al. 2019)

depth. In the second case, there could be some problem with the pressure switch. This device has the role of measuring system pressure and interrupting its operation if the pressure is not adequate. If the pressure switch is not working properly, the center pivot may operate at inadequate pressures, masking possible problems in system pressurization.

Figure 10.7b shows a circular pattern of low vegetative vigor, that is, there is a problem in some emitters located on the spans of the middle section of the center pivot. In this case, it can be inferred that some emitters need to be replaced or their deflector plates and/or nozzles need to be repaired. Depending on the crop, the circular pattern shown in Fig. 10.7b is an event that will hardly be noticed in an in loco monitoring, since the problem is located in the middle section of the center pivot and access to it is limited by the canopy of the crop, if it is tall.

The detections of problems in irrigated systems mentioned in this topic are some of the potentials that satellite and UAV images can offer to assist in monitoring and increase the efficiency of irrigated production systems. These technologies bring many other benefits to the digital irrigation industry and digital agriculture as a whole. Monitoring with images, if well used in crop management, can lead to significant improvements in profitability and yield for the agricultural sector.

Finally, digital irrigation presents a set of techniques and tools capable of promoting a more rational use of water, which can lead to savings of water and electricity, as well as improvements in the quantity and quality of production. The main bases of digital irrigation are the consideration of the spatial variability of the water demand of crops and the technological innovations present in irrigation systems, allowing them to be used with high levels of automation. The use of images from UAVs and satellites and the use of data science and machine learning algorithm are also an important part of digital irrigation, which enables more strategic decision-making. Finally, it is worth highlighting that digital irrigation is a recent study area with high potential for improving irrigated agriculture. However, some technologies still need to be better developed for a more effective utilization of the technique.

Abbreviations/Definitions

- Field Capacity (FC): It is the moisture when the drainage flux in the soil profile becomes negligible, around 1 mm d^{-1}.
- Crop Coefficient (Kc): It is the relation between crop evapotranspiration under standard conditions and reference evapotranspiration.
- Water Stress Coefficient (Ks): It is the correction of crop evapotranspiration coefficient imposed by soil moisture.
- Soil Bulk Density (Ds): It is the ratio between the mass of soil and the volume that contains it.
- Crop evapotranspiration (ETc): It is the amount of water lost by a given crop to the atmosphere under normal growing conditions.
- Reference Evapotranspiration (ETo): Evapotranspiration of a Hypothetical Crop Grown without Water Restriction, which Has a Height of 12 Cm, Surface Resistance of 70 S M^{-1} and an Albedo of 0.23
- Hypoxia: Condition of low oxygen concentration, which will limit the plant development.
- Permanent Wilting Point (PWP): It is the moisture from which the plant cannot extract water from the soil.
- Irrigation Interval: It is the interval between two successive irrigations.

Take Home Message/Key Points

- Digital irrigation is a segment of digital agriculture that focuses on the management and monitoring of irrigated agriculture.
- Precision irrigation and digital irrigation refer to different concepts.
- Correct management of irrigation promotes water savings and yield gains.
- Irrigation digitization brings convenience and safety by collecting and interpreting data that can support the proper management of irrigated agriculture.
- Digital transformation is currently permeating all aspects of agriculture to produce more sustainably and to allow better farming management.

References

Alface AB, Pereira SB, Filgueiras R, Cunha FF (2019) Sugarcane spatial-temporal monitoring and crop coefficient estimation through NDVI. Revista Brasileira de Engenharia Agrícola e Ambiental 23:330–335

Dias SHB (2018) Evapotranspiração de referência para projeto de irrigação no Brasil utilizando o produto MOD16. 2018. 75 f. Dissertação (Mestrado em Engenharia Agrícola) – Universidade Federal de Viçosa, Viçosa, MG

Gavioli A, Souza EG, Bazzi CL, Schenatto K, Betzek NM (2019) Identification of management zones in precision agriculture: an evaluation of alternative cluster analysis methods. Biosyst Eng 181:86–102

Oliveira JT, Cunha FF, Oliveira RA (2018) Irrigação maximizada. Cultivar Máquinas 190:9–11

Santos RA, Mantovani EC, Filgueiras R, Fernandes-Filho EI, Silva ACB, Venancio LP (2020) Actual evapotranspiration and biomass of maize from a red–green–near-infrared (RGNIR) sensor on board an unmanned aerial vehicle (UAV). Water 12:2359

Chapter 11
Digital Livestock Farming

Mario L. Chizzotti, Fernanda H. M. Chizzotti, Gutierrez J. de F. Assis, and Igor L. Bretas

11.1 Introduction

The United Nations (UN) has established the Sustainable Development Goals (SDGs) to ensure a more prosperous and healthier planet. Some of the objectives are: to end hunger and all forms of malnutrition; to double the agricultural productivity and income of small producers to ensure sustainable food production systems; and to increase investment in rural infrastructure, research and technology development by 2030. From the requirement to increase production efficiency, the concepts of precision livestock farming and digital livestock farming have emerged and are based on the use of technology to monitor and manage the production system efficiently. The fundamental concept of precision agriculture is the management of spatial variability in the field of production. Similarly, precision livestock farming aims to manage the variability of each animal or a group of animals using software, sensors, actuators, and robotics and to collect, store, and process data that provide information for decision-making.

Each animal production system demands different types of solutions in digital livestock farming. Digital livestock farming has allowed ever greater control of the production system, transforming the management procedure of rural companies. In this context, efficient management depends on the collection of information and the ability to interpret it. The more data are gathered, the better the diagnosis of the efficiency of the activity being developed, and the greater the potential economic return.

M. L. Chizzotti · F. H. M. Chizzotti · G. J. de F. Assis
Department of Animal Science, Universidade Federal de Viçosa, Viçosa, Minas Gerais, Brazil
e-mail: mariochizzotti@ufv.br; fernanda.chizzotti@ufv.br; gutierrez.assis@ufv.br

I. L. Bretas (✉)
North Florida Research and Education Center, University of Florida, Marianna, FL, USA
e-mail: igor.bretas@ufv.br

© The Author(s), under exclusive license to Springer Nature
Switzerland AG 2022
D. Marçal de Queiroz et al. (eds.), *Digital Agriculture*,
https://doi.org/10.1007/978-3-031-14533-9_11

This chapter aims to introduce some technologies already in use or under development for digital livestock farming and to describe the advantages they can offer for monitoring and managing animal performance.

11.2   Remote Sensing in Pastures

Grazing land accounts for most of the world's agricultural land and covers about one-fourth of the earth's land area. A significant amount of meat comes from extensive farming systems, with pasture being the basis of ruminant feed. Thus, the productivity of livestock is directly connected to the efficient management of pastures, in order to meet the nutritional demand of the animals and to preserve the ecosystem they inhabit. In this context, "digital grazing management," by which remote sensing is used to monitor the quantity and quality of pasture, as well as its botanical and structural composition, has been developed for monitoring of grasslands.

The estimation of an area's forage biomass is important for adjusting stocking rates, as well as for determining the grazing management of paddocks, which support pasture productivity. Moreover, in extensive farming systems pasture is the main source of energy for animals. The knowledge of the mass and nutritive value of forage available in an area allows the formulation of specific supplements in order to precisely meet the animals' nutritional requirements. The traditional methodologies used to obtain this information are based on the collection of samples from the pasture, which are weighed and sent for laboratory analysis of the forage. This process is extremely laborious, time-consuming, and expensive. In addition, pasture areas tend to have huge spatial variability, related to nutritional aspects and to botanical and structural composition. This undermines the representativeness of the samples, especially in areas that are large or hard to reach, which in turn reduces the accuracy of the data collected and, consequently, the quality of the information obtained.

Technological advances in the agricultural sector have allowed producers to minimize the problems mentioned above, making their activities more profitable. Currently, many technologies have already become used in agriculture with the implementation of so-called digital agriculture. However, in the livestock sector, the implementation of technological packages for pasture management and monitoring still faces resistance from researchers, technicians, and producers and has not been widely explored in many countries. The Australian "Pastures from Space" project (https://agric.wa.gov.au/n/7700) is one of the most prominent examples of innovation focused on pastures and has developed tools to provide information on biomass, growth rate, and quality of forage almost in real time at the farm level, using moderate- to high-resolution satellites for remote sensing.

Remote sensing is an important tool that can be used in the management and monitoring of pasture areas following the same principles already used in agriculture, that is, by using remote images to obtain estimates of the available forage biomass, the nutritive value of the forage, the botanical composition of the area, and the spatial distribution of forage. Remote sensing, whether terrestrial (using proximal sensors), suborbital (using sensors on unmanned aerial vehicles), or orbital (using sensors on satellites), allows the collection of more data in less time, generating more information and with greater precision. In addition, it allows the monitoring of larger areas that are difficult to access.

The cost of acquiring information for pasture management is usually low, especially when using images generated by sensors coupled to satellites, as with images from the Landsat and Sentinel satellites, that are made available for free. These satellites provide information with moderate spatial resolution and good temporal resolution, for monitoring pastures.

With satellite images, using mainly the infrared and red bands, one can determine the normalized difference vegetation index (NDVI), which shows a high degree of correlation with the biophysical properties of pastures and the percentage of green cover in an area (Ferreira et al. 2013). Some companies now offer NDVI maps directly on mobile devices so that decision-making information is now quickly accessible and at an affordable cost.

Some applications for remote sensing in the field of animal science will be presented below. However, before we discuss some of the main practical applications of remote sensing in pastures, it is worth mentioning that the first step for the application of remote sensing in pasture management is calibration, as this is an indirect method for estimating the biophysical parameters of vegetation. It is therefore necessary to obtain a database using a direct method to build a mathematical model capable of estimating such parameters with data obtained by an indirect method, for example, by using NDVI to estimate biomass.

11.3 Stocking Rate Adjustments and Grazing Management

In extensive farming systems involving either continuous or intermittent stocking, the number of animal units per hectare (AU/ha) must be determined based on the animals' dry matter demand, which varies according to animal weight and category, and the availability of forage in the paddock. To determine the availability of forage, images acquired by satellites or unmanned aerial vehicles (UAVs) can be used.

The main problem with the use of satellite images is the presence of cloud cover, which makes it difficult to obtain images at certain periods. Hence, temporal resolution can be limited in this case. In addition, the downloading and processing of satellite images can require high computational power and be very time-consuming. Currently, the development of cloud-processing techniques allows this problem to be circumvented through computational resource-sharing services. One example of this is the open-source platform Google Earth Engine, which allows the processing of satellite images and geospatial data over a large timescale entirely in the cloud.

To solve problems related to cloud coverage in images, in addition to seeking greater temporal and spatial resolution, multispectral digital cameras coupled to UAVs can be used. This type of system requires the use of specific software for processing the information generated, as well as greater computational resources and trained personnel. This type of system undoubtedly generates information of the highest level, but at a higher cost per hectare. It is therefore important to make a prior analysis of the use of technologies. The use of satellite images may differ depending on the circumstances and needs.

Among the costs involved in the use of UAVs, multispectral cameras (the sensors) can vary widely in terms of price and model. Some models cost over US $5000. One option is to use conventional cameras, which cost less. These cameras usually have a single sensor and use RGB bands by default. Some of them, costing around US $500, can be modified to make the NDVI calculation feasible. In some studies, images obtained by cameras coupled to UAVs showed a good correlation with the height and biomass of the pasture (Borra-Serrano et al. 2019; Batistoti et al. 2019). However, it should be emphasized that the use of low-cost cameras requires greater care in terms of calibration.

The vegetation indices are linear transformations of two or more spectral bands, mainly in the red (R) and near-infrared (IV) bands of the electromagnetic spectrum. These indices are dimensionless, and they correlate with biophysical parameters of vegetation, such as biomass, leaf area index (IAF), green vegetation vigor, chlorophyll content, and photosynthetic activity and productivity (Pezzopane et al. 2019). Images used for remote sensing applied to crops or pastures, mainly satellite images, may be affected by topographic or atmospheric interference, cloud cover, or the influence of exposed soil. Therefore, the use of these indices is based on the fact that the implementation of two or more spectral bands can minimize the main sources of "noise" that affect the responses of the vegetation, as they are more sensitive than the individual spectral bands when related to biophysical vegetation parameters. Table 11.1 shows some indexes that can be applied to pasture monitoring.

It is important that care be taken when choosing the type of sensor or camera to be used, as the different types of sensors have different spectral, radiometric, spatial, and temporal resolutions. Therefore, when it comes to tropical pasture areas, the temporal resolution of the image becomes highly relevant, as during the growing

Table 11.1 Indexes related to biophysical parameters of vegetation

Index	Equation	References
SRVI	NIR/RED	Jordan (1969)
NDVI	(NIR − RED)/(NIR + RED)	Rouse et al. (1974)
SAVI	1.5 (NIR − RED)/(NIR + RED + 0.5)	Huete (1988)
NDRE	(NIR − RE)/(NIR + RE)	Gitelson and Merzlyak (1994)
OSAVI	(NIR − RED)/(NIR + RED + 0.16)	Rondeaux et al. (1996)
GNDVI	(NIR − GREEN)/(NIR + GREEN)	Gitelson et al. (1996)
EVI	2.5*((NIR − RED)/(NIR + 6 RED − 7.5 BLUE + 1))	Huete et al. (1997)

NIR, RED, RE, GREEN, and BLUE refer to the reflectance in the near-infrared, red, red edge, green, and blue wavelengths, respectively

period of the grass, large structural and biomass variations occur in a short time, so the use of satellites with very long revisitation intervals is not recommended. The Landsat satellite, for example, will repeat the image in the same area every 16 days. In some cases, the Sentinel satellite can be more advantageous, since its temporal resolution is every 5 days. Similarly, small paddocks cannot be monitored with low spatial resolution images, since the pixel size often exceeds the dimensions of paddocks, leading to a loss of representativeness.

Bretas et al. (2021) used artificial intelligence algorithms to predict the biomass of pastures formed by *Brachiaria sp.* (syn. *Urochloa*), as well as their dry matter concentration, by combining vegetation indices obtained by satellites and meteorological data. They demonstrated that, beyond its application in agriculture, remote sensing enables great advances in the monitoring and management of pastures. In addition to adjustments of the stocking rate through estimates of available biomass, the vegetation indices obtained with these images can also be used for grazing management, that is, they allow the producer to manage the forage harvest of the animals.

The NDVI, the main vegetation index used to monitor vegetation, varies from -1 to 1: the higher the index value, the greater the biomass available. In general, watercourses present negative NDVI values, exposed areas of soil are close to zero, and areas with vigorous vegetation show NDVI values above 0.6. Suppose that a livestock farm with paddocks is being used for grazing in an intermittent stocking system. It is known that control of the moment of the animals' entry and exit must be carried out according to the canopy height target, which will vary according to the forage type used. Thus, knowing the correlation between the height of the forage and the vegetation index, an NDVI map of the area can be used to identify which paddocks are at the animals' moment of entry and which ones are at their moment of exit. Machine-learning techniques, discussed in another chapter of this book, such as the use of a supervised classification algorithm, can be used to indicate when to start or stop grazing in certain pasture areas, making decision-making faster and more accurate.

Hypothetically, paddocks with an average NDVI above 0.7 will have dense vegetation, indicating that this is the moment to start grazing. Paddocks with an average NDVI below 0.3 will be at the animals' moment of exit, while those with intermediate NDVI values will be in the process of regrowth. Thus, NVDI mapping has the potential to assist the grazing management to achieve a sustainable pasture carrying capacity, optimizing the use of the area and avoiding overgrazing that can be prejudicial to the persistence of the pasture. Other indices can be combined with the NDVI to generate more accurate estimates and maps, since in conditions of very dense vegetation, the NDVI index saturates and loses sensitivity to changes in vegetation biomass (Gu et al. 2013). Figure 11.1 illustrates the temporal variation of NDVI in an area under grazing, and demonstrates how maps of vegetation indexes can assist decision-making in the field.

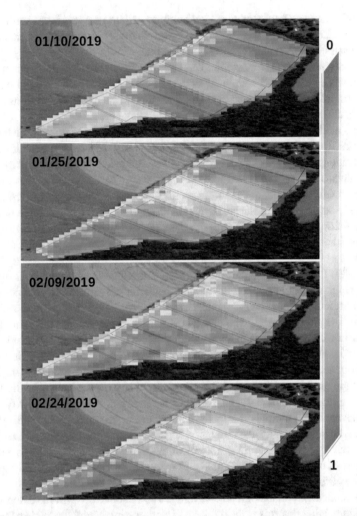

Fig. 11.1 NDVI maps obtained by satellite images in paddocks managed under intermittent stocking, at different times in the grazing cycle. The paddocks with NDVI closer to 1 indicate high forage biomass and possibility of entry of the animals, while paddocks with NDVI closer to 0 indicate bare soil or overgrazed areas. Intermediate NDVI values indicate moderate forage biomass, characterizing the moment to move animals for another paddock or a post-grazing regrowth period

11.4 Weed Detection and Botanical Classification

In addition to providing estimates of forage biomass and identifying the appropriate moment to start or stop grazing in pasture areas, the remote sensing of pastures allows automated detection of weeds in the area, as well as the determination of botanical composition in areas of livestock farming integration or intercropping between grasses and legumes. This information is relevant for the quick and efficient control of undesirable plants, the quantification of the above-ground biomass that is actually represented by pasture, or the quantification of the proportion of legumes in a pasture area, which directly impact the management of grazing and nitrogen fertilization. In tropical pastures, for example, the participation of legumes is expected to be 20–40% of the forage mass, and the use of images obtained by RGB or multispectral cameras on UAVs could allow this estimate to be made quickly and accurately. In addition, weed control is essential for maintaining the productivity and longevity of a pasture. It is known that the control of weeds in the field is time-consuming, laborious, and costly. In addition, the use of herbicides applied indiscriminately over a total area can cause a great environmental impact. Thus, automated weed identification would enable timely control to prevent productive and economic damage, as well as enable the rational use of chemical herbicides.

Artificial intelligence algorithms, such as deep learning techniques, can be trained to recognize each of the components within a pasture from images obtained by remote sensing. In this way, UAV sprayers equipped with global positioning systems (GPS) can be directed to specific points where there is a need to control weeds, minimizing operating time, herbicide costs, and environmental impact. Valente et al. (2019) and Petrich et al. (2020) demonstrated high accuracy in the detection of toxic plants in areas under pasture by using aerial images captured by RGB cameras on UAVs, with machine-learning techniques. Figure 11.2 shows an image obtained by a low-cost RGB camera coupled to a UAV in pasture areas and demonstrates high spatial resolution for the identification of weeds or components with no value for animal feeding in the area.

11.5 Forage Nutritive Value and Animal Nutrition

The monitoring of pastures by remote sensing, which forms part of the concept of "digital livestock farming," also supports advances in the development of new supplements for animal nutrition. This is possible because the quality of pasture, in terms of its nutritional value, can be estimated from the vegetation indexes generated through digital images. It is possible to estimate the content of crude protein, neutral detergent fiber (NDF), acid detergent fiber (ADF), and digestibility, among other forage parameters, in an area of interest and thereby formulate an adequate supplement for this situation. Wijesingha et al. (2020) used a hyperspectral camera on a UAV to estimate the crude protein and FDA content of eight different pasture areas with high variability of botanical composition and with different cutting regimes, thus demonstrating that it is possible to estimate the quality of forage with precision, regardless of the type of pasture.

The prediction of the nutritional composition of pasture is fundamental for the formulation of feed supplements in order to optimize animal growth. In the event of a high growth rate of forage due to favorable climatic conditions during a rainy season, an imbalance may occur in animal diets if the high energy availability of the

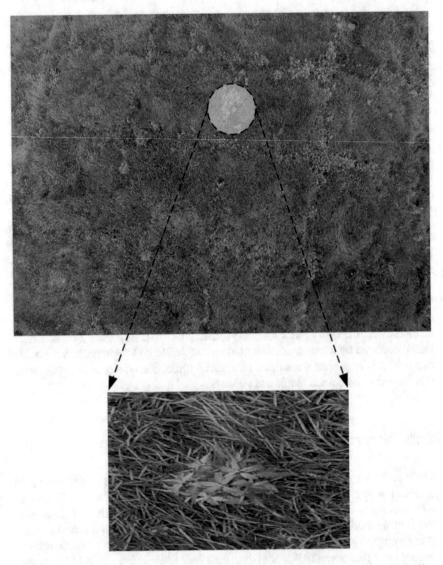

Fig. 11.2 RGB image, obtained by a UAV flying at a height of 35 m in a pasture area, highlighting a weed species

pasture is not accompanied by a proportionate increase in the availability of crude protein. It is therefore often necessary to supplement the animals' intake of crude protein. For the formulation of such a supplement, it is essential to know the amount of crude protein that the pasture will be unable to provide, which should therefore be provided by the supplement. To this end, remote sensing can be employed to obtain the protein content of the pasture at a low cost and in a timely manner in order to determine the ideal level of protein in the supplement. The principle applied in pasture areas is the same used for agricultural crops in the development of precision agriculture. Vegetation indices (NDVI, GNDVI, EVI, among others) are used, which correlate with the chlorophyll concentration in the plant. As most of the nitrogen present in plants is found in chlorophyll, it is correlated with the nitrogen concentration in the plants. Consequently, it is possible to estimate the crude protein content of the forage from images obtained by remote sensing and, thus, formulate a concentrated supplement for each specific area and period of the grazing cycle.

11.6 Automated Systems for Feeding

Feed often represents the most significant component of livestock production costs, and monitoring them is crucial for optimizing returns. Electronic feeders have been developed for several species, allowing continuous and automatic monitoring of animals to support production decisions at the farm level and deliver specific diets for each animal or barn.

Feed intake monitoring can be used for adjusting the diet for the optimal nutrient requirements, thus decreasing environmental impact, or for the early detection of disease, thus decreasing the use of antibiotics and avoiding the spread of infectious diseases (Pomar and Remus 2019). These systems consist of feeders supported on load cells, which enable the electronic recording of the total amount of feed consumed by each animal. The weight difference recorded by the feeders between the entry and exit of the animal is understood to be the feed intake. Identification of individual animals is made by means of an electronic tag, usually implanted in the ear of each animal, and detected by radio frequency identifier (RFID) readers at the feeders. At each animal's visit, data are recorded that can be automatically extracted, including daily intake (kg/day), feed intake per visit, intake rate (grams/minute), time spent in the feeders without consumption, intervals between meals, and hierarchy, allowing the interpretation of animal behavior or health issues as well as improving the prediction of animal performance and feed efficiency of individuals or group of animals. Figure 11.3 shows electronic feed bins (Intergado Systems) at the Department of Animal Science in UFV, Brazil.

In addition to the monitoring of feed intake, feeding systems that automatically load, mix, and deliver feed to the animals have also been developed. The ingredients can be pulled from the feed mill or silos, mixed in the mixer, and then delivered to the feedbunk using automatic feeding robots that drive on wheels or rails to distribute feed, which can increase work efficiency on the farm. The software and precision in the automated feeding systems deliver a different ration to different barns or animal categories, with adjusted amounts of energy, protein, mineral, and additives, improving the feed efficiency and decreasing the excretion of nutrients by matching the nutritional requirements of the animals with the delivered diet. The system also allows multiple and smaller feedings throughout the day, maximizing the dry matter intake and decreasing the feed sorting and feed wasting.

Fig. 11.3 Electronic feed bins (Intergado Systems) at the Department of Animal Science in UFV

Computerized calf feeders are also already in use to precisely deliver instant portions of fresh milk at a controlled temperature, according to each calf's feeding plan, allowing the reduction in labor costs and improving the health of calves which reduces veterinary costs in the long term.

11.7   Automated Systems for Body Weighing

Monitoring the body weight change of growing animals is important to keep track of their health status and determine the best time for their mating or slaughter. One of the difficulties of traditional weighing methods is the need to lead the animals to the scale, which results in stress and increases labor costs. Solutions are available in the market to overcome these obstacles, including the walking over scale, which records the animal as it moves through the weighing platform. In this system, every time an animal passes over the scale, its weight is measured, allowing the producer to know the daily weight gain of individual cattle and the entire herd. The scale works remotely in pastures or feedlots using energy supplied by solar panels. Another technology involves scales installed in front of the drinking fountain in the animals' pens, weighing the animals whenever they drink water with the advantage of a more static weighing that improves the accuracy of the measurements.

The continuous monitoring of the body weight can be useful to describe and model the growth curve of each animal. The animal's growth is curvilinear and s-shaped, known as sigmoid growth curve, a pattern of growth in which the rate of weight gain increases initially in a positive acceleration phase and then declines in a negative acceleration phase, until zero growth rate or mature/adult weight has been reached (Fig. 11.4).

This slowing of the rate of weight gain associated with animal's age or days-on-feeding decreases the daily carcass deposition, nonetheless, the dry matter intake is still increasing, which affects the feed conversion ratio and therefore the instant profitability of each animal or barn. In other words, the monitoring of the growth can be used to describe the instant income of each individual and compare it with the associated costs to decide the best moment to send them to the slaughterhouse, in order to maximize the profit.

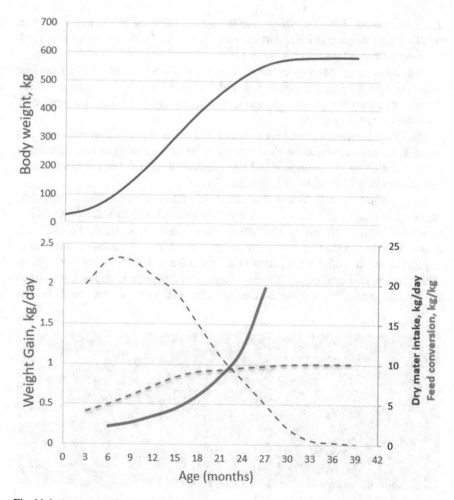

Fig. 11.4 An example S-shaped growth curve pattern estimated by autonomous weight scale (blue line) and the rate of weight gain (dashed blue), dry matter intake (dashed green), and feed conversion ratio (red), in bovine cattle

The monitoring of the growth curve can also be useful to predict the changes in the body composition of the animals and therefore the carcass composition to achieve the optimum fatness level and used in decision-making for meat production.

The continuous monitoring of growth is also helpful in the early detection of health issues, ambient stress, or management failures, by generating alerts for unexpected variations on the growth curve, allowing a quick response to correct the cause of the problem and therefore decreasing the associated economic losses.

11.8   Automated Voluntary Milking Systems

There are several different robotic milking systems already in use in dairy farms, enabling the reduction of labor and allowing 24 h/day of milking, which can maximize the milk yield (Fig. 11.5).

In those systems, cows enter the milking station, driven by the desire to relieve udder pressure to receive concentrated feed during milking, and after RFID identification, robotic arms sanitize the teats, and lasers then guide suction tubes onto each teat to start the milking.

The systems also enable the continuous monitoring of production and quality of the milk, as well as the real-time monitoring of the health and reproduction status of the cows, allowing the collection and interpretation of useful data for management decision-making to improve farm profitability.

The systems can test in real time the milk's fat content, protein content, and somatic cell counts, and if the milk is not appropriate, it can be instantly separated for calves' feeding, increasing substantially the marketed milk quality. Furthermore, it can detect progesterone levels for pregnancy or heat detection automatically improving the reproductive performance of the herd. The collected data during milking allows the evaluations of the performance at the individual level, which can be used to maximize the production and the efficiency of the herd, by the

Fig. 11.5 Examples of commercial robotic milking systems. (Source: https://upload.wikimedia. org/wikipedia/commons/4/41/Dairy_Campus_Ljouwert05.jpg; https://commons.wikimedia.org/ wiki/File:Melkrobot2.JPG)

identification of nonprofitable cows, or to promote improvement in the nutrition plan or in the environment of the cows.

The use of robotic milking systems also allows the automatic phenotyping of the herd, which can be used to estimate the breeding value for genetic selection of more efficient and healthier cows.

11.9   Wearable Animal Sensors

Wearable technologies and biosensors have been developed for livestock management practices. Such devices can provide real-time information about animal health status. Some of them are already in use in dairy and beef farms, such as RFID tags; GPS; accelerometers; thermometers; and intraruminal pressure, temperature, and pH boluses (Fig. 11.6).

11.10   Bolus pH and Temperature Meter

Adequate ruminal health promotes higher feed intake and digestibility, and, consequently, better animal performance. The adequate ruminal condition has been defined as a physiological ruminal pH varying from 5.8 to 6.8. Ruminal acidosis (pH lower than 5.8) decreases feed intake and milk production, reduces weight gain, and can cause other health problems such as laminitis, reducing animal productivity.

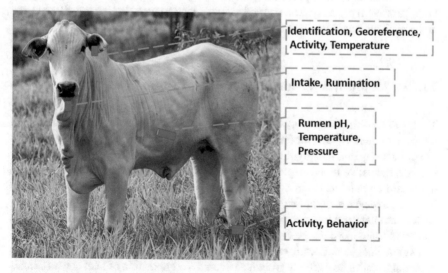

Fig. 11.6 Wearable sensors to monitor health status, location, activity, and behavior

The use of intraruminal boluses allows the monitoring of the temperature and pH of the rumen, and several commercial models are already on the market, differing mainly in terms of their lifespans, which vary from 2 months to 3 years. These sensors are orally inserted and are located inside the rumen and emit wireless information to readers installed at strategic locations in feeders or pens, generating real-time data about rumen conditions to support nutrition management. According to Jonsson et al. (2019), a minimum of nine animals with boluses would provide a reasonable estimate of the average pH of the herd or pen and would be a helpful tool for precise formulation of the diets in intensive systems.

11.11   Global Positioning System Collars

Solar-powered collars equipped with a GPS system enable individual tracking of the animals, which can be useful for monitoring their daily physical activity or the distance they walk to reach water sources or supplement feeders, for example, helping in the design of paddocks as well as for the estimation of the energy expenditures with physical activity for a more precise supplement formulation.

GPS collars can also be used to monitor the time spent under grazing of each area of the paddock and generate heat maps to identify over or under-grazed areas to improve the efficiency of use of the pastureland.

Other uses of GPS in livestock include virtual fencing, which enables livestock to be moved or restrained in a specified area. The equipment sends coordinate data to a terrestrial station, recording activities and positions in real time. The collar emits an audio tone or electrical pulse if the animal approaches the limits of the allowed geocoordinates, preventing access to a protected area or crossing property boundaries of a property. The warned animals tend to turn back to the allowed virtual area. This might be useful in extensive rangelands.

11.12   Computer Vision in Livestock Production

The rapid growth of digital technology, with the replacement of analog methods and the development of new algorithms, has led to a significant increase in the number of applications that use digital image processing (DIP). Accompanying this development, the use of image capture devices, such as three-dimensional (3D) cameras, has become available, enabling the extraction of biometric information contained in images. However, several factors influence the capture and quality of these images, such as ambient lighting and the position of the animal, and those factors might be corrected to improve the predictions.

Digital images collected with a camera positioned above the animal have the potential to estimate the body weight and carcass composition of beef cattle (Gomes et al. 2016). The use of images has the potential to replace the traditional weighing

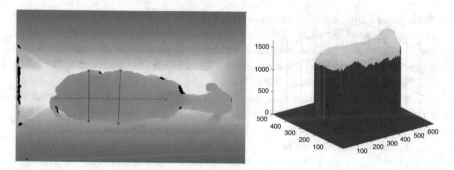

Fig. 11.7 Example of computer vision analysis of 2D (left) or 3D (right) images in beef cattle

of cattle since the images can be obtained automatically in the environment in which the animals are kept. This type of system uses digital image processing to evaluate biometric measurements to predict the weight of animals. The most common biometric measurements include body area, body volume withers and hip height, body length, hip width, and chest girth. The automation of the data capture and processing of the images to predict the weight of the animals can soon enable the real-time monitoring of body growth throughout video image analysis (VIA). This method may replace the traditional weighing on scales or the use of ultrasound to predict the fatness of the carcasses, since it has the advantage of not requiring any handling of the animals. Figure 11.7 shows an example of computer vision analysis of 2D (left) or 3D (right) images in beef cattle.

Computer vision and AI systems can also be used for surveillance or behavior monitoring of animals in the pen, identifying the daily activity or the time spent in feeding or drinking areas, monitoring feeding behavior, or detecting irregular patterns that might be related to disease or environmental stress. Deep Learning algorithms has been applied to anomaly detection in swine farms, translating video images into information about a pig's aggressiveness patterns to support swine wellbeing. In poultry, VIA has been used to monitor flock motion patterns, floor occupied area by the birds, and the feeding behavior; to estimate the level of animals' welfare in response to stocking density and environmental conditions; and to improve health, welfare, and efficiency in livestock production.

11.13 Infrared Thermography in Livestock Production

In animal production, the use of new technologies such as infrared thermography has emerged, among other applications, as an alternative for clarifying the impact of environmental factors in decision-making and the promotion of animal health and wellbeing (Roberto and Souza 2014). Infrared thermography involves measuring infrared energy using cameras that capture the radiation emitted by a specific body. A thermographic camera detects temperature variations, ranging from hot to cold, which in turn indicate physiological abnormalities in animals (Turner and Eddy 2001; Infernuso et al. 2010). This device is of great value for observing the stress levels in farm animals. In livestock, studies have shown the potential use of thermographic images to evaluate the thermal comfort of swine (Kotrbacek and Nau 1985) and poultry (Nascimento et al. 2011).

The thermographic images are also effective for the detection of inflammatory processes, such as laminitis or mastitis in lactating dairy cows, and the detection of chronic pain after the tail docking in swine (Eicher et al. 2006) or fever of animals, with several application on monitoring of the health status of the herd. Infrared cameras can verify body parts' temperature without physical contact to alert farmers about the health or stress issues.

Infrared thermography has also the potential to predict the metabolic heat losses of the animals. Gomes et al. (2016) found that the ocular surface temperature was positively correlated to the metabolic heat production in beef cattle, and that ocular temperature was more accurate than skin surface temperature in the evaluation of the heat losses of the animals.

Another application of infrared thermography is the identification of animals predisposed to variation in meat quality in response to stress suffered before slaughter (Tong et al. 1995; Schaefer et al. 2001, 2018). Infrared thermography of carcasses also has the potential to be used in predicting the fatness, as well as in the early identification of carcasses that will produce less tender meat, which enables infrared thermography as a potential tool for carcass grading protocols. Figure 11.8 shows examples of infrared thermography images in cattle or beef carcasses.

11.14 Digital Carcass Grading

The objective of carcass classification is to draw up a hierarchy of carcasses based on clearly defined quality attributes to guarantee meat quality and, consequently, consumer satisfaction. The classification is based on the description of several

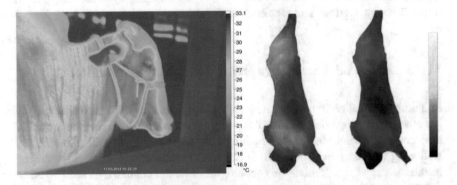

Fig. 11.8 Examples of infrared thermography images in cattle or beef carcasses

intrinsic and extrinsic factors that affect the composition and quality of the carcasses, with consequent effects on the quality of the meat produced. The composition and quality of the carcass vary according to species, age, type of maturity, sex class, and the effects of interaction with production systems and technologies (Webb and Erasmus 2013). These characteristics make it possible to classify carcasses effectively into categories so that carcasses of similar quality are classified in the same category, reducing the variation between carcasses and in the meat produced.

The objective measurement, in real time, of the carcass characteristics that determine the final quality of the meat is an important development for the industry, and it is progressing mainly due to digital image analysis technologies. Some of these technologies are still far from applicable to the slaughterhouse environment, while others are already well advanced and in the process of commercialization or are already being used in meat industries worldwide. The technologies used include high-definition cameras, computed tomography, X-ray imaging, among other systems.

VIA devices have been developed for the European Union classification system (SEUROP), capturing digital images of half carcasses before refrigeration, allowing the measurement of lengths, heights, angles, areas, and volumes and the consequent classification of the carcass fatness and degree of conformation. Examples of such systems are VBS 2000 (E + V technology GmbH and Co.KG), from Germany, and BCC-2 (Carometec A/S), from Denmark. VIA can also be used to detect the distribution and proportion of the different tissues covering the surface of the carcass in order to estimate the degree of fatness (Fig. 11.9).

The second type of application of VIA for carcass grading involves obtaining images from a cross section of the *Longissimus* muscle, taken between the 12th and 13th ribs of the bovine carcass. The image analysis enables estimation of meat yield, through analysis of the rib eye area and measurement of subcutaneous fat thickness, and of meat quality, through the color analysis of the muscular tissue and the degree of marbling. There are already instruments available on the market, such as VIAScan from Australia; the VGB 2000 (E + V technology GmbH and Co.KG) from Germany; and CVS (Computer Vision Systems) from Canada. The Meat Image Japan (MIJ) camera has been used to objectively predict the degree of marbling, rib eye area, and other characteristics that the Japanese industry values when evaluating the carcasses of Wagyu cattle in Japan.

Hyperspectral or multispectral cameras are also being evaluated as a promising alternative for nondestructive quality assessment of meat (Cluff et al. 2008). Hyperspectral images have already been applied to meat products for the classification of pork meat quality (Qiao et al. 2007), the detection of tumors in chicken carcasses (Park et al. 2007; Nakariyakul et al. 2008), fish quality assessment (Sivertsen et al. 2009), and beef tenderness assessment (Konda Naganathan et al. 2008).

The evaluation and control of carcass quality is still carried out visually and subjectively in slaughter plants, which is expensive, time-consuming, and subject to bias. The meat industry can benefit from the use of emerging technologies that performs objective, non-destructive, fast, and accurate methods for the evaluation of

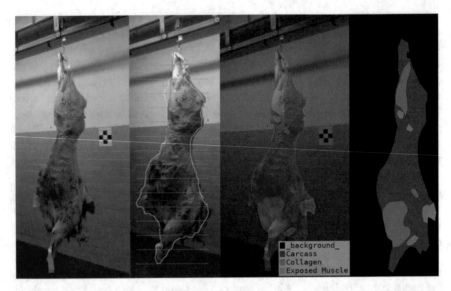

Fig. 11.9 Example of VIA with biometric measurements and the identification of subcutaneous fat (red area), exposed muscle (yellow area), or connective tissue (green area) using AI system in beef carcass

carcasses and meat quality. The development and adoption of digital quality assessment systems may produce economic benefits to the meat industry, by increasing consumer confidence in the quality and standard of meat products.

11.15   Final Remarks

The use of sensors in the animal production area has enormous potential, and its development will promote both production efficiency and product quality. Digital livestock farming will enable an extraordinary flow of information collected in real time, generating an almost unlimited number of possibilities for control and intervention that are not feasible using traditional production systems. The application of these technologies can generate major changes in the animal production chain with the creation of new service sectors, greater efficiency in the use of resources, less environmental impact, and greater added value for products of animal origin.

Abbreviations/Definitions

- Computer vision systems (CVS): Consists of the development of artificial systems to deal with visual problems of interest by means of image processing and analysis techniques, as well as machine learning and pattern recognition. In summary, the method is based on the use of one or more cameras to acquire images or videos of live animals or carcasses.

- Precision livestock farming: Can be defined as the use of technologies to monitor the variability between animals, or groups of animals, and the environment that are inserted on a large scale, continuously and at a low cost to optimize decision-making and guarantee greater productive efficiency, animal welfare, sustainability, and profitability.
- Remote sensing: Can be defined as the use of sensors to collect data at a distance, without the need for contact. In the case of remote sensing of pastures through drones and satellites, the term can be understood as the imaging of the pasture from a distance.
- Vegetation indexes: Are linear combinations between two or more bands of the electromagnetic spectrum that are used to correlate with variables of interest in the pasture. These indices are dimensionless and are used because they are more sensitive to changes in vegetation compared to the use of isolated bands, due to possible atmospheric, soil, and relief interferences, among other noise sources.
- Wearable sensors: Sensors integrated into wearable objects or directly contact with the animal body in order to help monitor health, behavior, and nutrition and provide reliable data for improving the management of animals.

Take Home Message/Key Points

- The rising demand for animal products and the growing concern about issues related to animal welfare, food safety, and sustainability drive improvements in the efficiency of animal production.
- Precision livestock farming technologies can offer advantages for automatization of pastures and animal management and decision-making support.
- The use of sensors in the animal production area has enormous potential, and its development will promote both production efficiency and product quality.

References

Batistoti J, Junior M, Matsubara E, Filho GS, Akiyama T, Gonçalves W, Liesenberg V (2019) Estimating pasture biomass and canopy height in Brazilian savanna using UAV photogrammetry 1-12. Remote Sens 11:2447. https://doi.org/10.3390/rs11202447

Borra-Serrano I, de Swaef T, Muylle H, Nuyttens D, Vangeyte J, Mertens K, Saeys W, Somers B, Roldán-Ruiz I, Lootens P (2019) Canopy height measurements and non-destructive biomass estimation of *Lolium perenne* swards using UAV imagery. Grass Forage Sci 74:356–369. https://doi.org/10.1111/gfs.12439

Bretas IL, Valente DSM, Silva FF, Chizzotti ML, Paulino MF, D'Áurea AP, Paciullo DSC, Pedreira BC, Chizzotti FHM (2021) Prediction of aboveground biomass and dry-matter content in *brachiaria* pastures by combining meteorological data and satellite imagery. Grass Forage Sci:gfs.12517. https://doi.org/10.1111/gfs.12517

Cluff, K., Naganathan, G. K., Subbiah, J., Lu, R., Calkins, C. R., & Samal, A. (2008). Optical scattering in beef steak to predict tenderness using hyperspectral imaging in the VIS-NIR region. Sensing and Instrumentation for Food Quality and Safety, 2(3), 189–196. https://doi.org/10.1007/s11694-008-9052-2

Eicher SD, Cheng HW, Sorrells AD, Schutz MM (2006) Behavioral and physiological indicators of sensitivity or chronic pain following tail docking. Short communication. J Dairy Sci 89:3047–3051. https://doi.org/10.3168/jds.S0022-0302(06)72578-4

Ferreira L, Fernandez L, Sano E, Field C, Sousa S, Arantes A, Araújo F (2013) Biophysical properties of cultivated pastures in the Brazilian savanna biome: an analysis in the spatial-temporal domains based on ground and satellite data. Remote Sens 5(1):307–326. https://doi.org/10.3390/rs5010307

Gitelson, A. & Merzlyak, M. N. (1994) Quantitative Estimation of Chlorophyll-a using Reflectance Spectra: Experiments with Autumn Chestnut and Maple Leaves. 6. J. Photochemistry and Photobiology B: Biology, 22, 247–252. https://doi.org/10.1016/1011-1344(93)06963-4

Gitelson, Y.j. Kaufman, M.n. Merzlyak. (1996) Use of a green channel in remote sensing of global vegetation from EOS-MODIS. Remote Sens. Environ., 58 (3), pp. 289–298. https://doi.org/10.1016/S0034-4257(96)00072-7

Gomes RA, Monteiro GR, Assis GJF, Busato KC, Ladeira MM, Chizzotti ML (2016) Technical note: estimating body weight and body composition of beef cattle through digital image analysis. J Anim Sci 94(12):5414–5422. https://doi.org/10.2527/jas.2016-0797

Gu Y, Wylie BK, Howard DM, Phuyal KP, Ji L (2013) NDVI saturation adjustment: a new approach for improving cropland performance estimates in the Platte River basin, USA. Ecol Indic 30:1–6. Available in https://doi.org/10.1016/j.ecolind.2013.01.041

Huete, A. R. (1988) A soil-adjusted vegetation index (SAVI). Remote Sensing of Environment, 25 v.3, p.295–309, https://doi.org/10.1016/0034-4257(88)90106-X

Huete, A.R. et al. (1997) A comparison of vegetation indices over a global set of TM images for EOS-MODIS. Remote Sensing of Environment. v.59, n.3, p.440–451. https://doi.org/10.1016/S0034-4257(96)00112-5

Infernuso T, Loughin CA, Marino DJ, Umbaugh SE, Solt PS (2010) Thermal imaging of normal and cranial cruciate ligament-deficient stifles in dogs. Vet Surg 39:410–417. https://doi.org/10.1111/j.1532-950X.2010.00677.x

Jordan, C.F., (1969). Derivation of leaf-area index from quality of radiation on the forest floor. Ecology 50, 663–666. https://doi.org/10.2307/1936256

Jonsson NN, Kleen JL, Wallace RJ, Andonovic I, Michie C, Farish M, Mitchell M, Duthie C-A, Jensen DB, Denwood MJ (2019) Evaluation of reticulo ruminal pH measurements from individual cattle: sampling strategies for the assessment of herd status. Vet J 243:26–32. https://doi.org/10.1016/j.tvjl.2018.11.006

Konda Naganathan G, Grimes LM, Subbiah J, Calkins CR, Samal A, Meyer GE (2008) Visible/near-infrared hyperspectral imaging for beef tenderness prediction. Comput Electron Agric 64:225–233. https://doi.org/10.1016/j.compag.2008.05.020

Kotrbacek V, Nau HR (1985) The changes in skin temperatures of periparturient sows. Acta Vet Brno 54:35–40. https://doi.org/10.2754/avb198554010035

Nakariyakul, S., & Casasent, D. (2008). Hyperspectral waveband selection for contaminant detection on poultry carcasses. Optical Engineering, 47, 087202–087209. https://doi.org/10.1117/1.2968693

Nascimento GR, Pereira DF, Nääs IA, Rodrigues LHA (2011) Índice fuzzy de conforto térmico para frangos de corte. Engenharia Agrícola 31:219–229. https://doi.org/10.1590/S0100-69162011000200002

Park, B., Windham, W. R., Lawrence, K. C., & Smith, D. (2007). Contaminant classification of poultry hyperspectral imagery using a spectral angle mapper algorithm. Biosystems Engineering, 96, 323–333. https://doi.org/10.1016/j.biosystemseng.2006.11.012

Petrich L, Lohrmann G, Neumann M, Martin F, Frey A, Stoll A, Schmidt V (2020) Detection of Colchicum autumnale in drone images, using a machine-learning approach. Precis Agric 21(6):1291–1303. https://doi.org/10.1007/s11119-020-09721-7

Pezzopane JRM, Bernardi AC, De C, Bosi C, Crippa PH, Santos PM, Nardachione EC (2019) Assessment of Piatã palisadegrass forage mass in integrated livestock production systems using a proximal canopy reflectance sensor. Eur J Agron 103:130–139. https://doi.org/10.1016/j.eja.2018.12.005

Pomar C, Remus A (2019) Precision pig feeding: a breakthrough toward sustainability. Anim Front 9:52–59. https://doi.org/10.1093/af/vfz006

Qiao, J., Ngadi, M.O., Wang, N., Gariépy, C., & Prasher S.O. (2007). Pork quality and marbling level assessment using a hyperspectral imaging system. *Journal of Food Engineering.* 83(1):10–16. https://doi.org/10.1016/j.jfoodeng.2007.02.038

Roberto, J. V. B.; Souza, B. B. Utilização da termografia de infravermelho na medicina veterinária e na produção animal. Journal of Animal Behaviour and Biometeorology, v. 2, n. 3, p. 73–84, 2014. https://doi.org/10.14269/2318-1265/jabb.v2n3p73-84.

Rondeaux, G., Steven, M., & Baret, F. (1996). Optimization of soil-adjusted vegetation indices. *Remote Sensing of Environment,* 55(2), 95–107. https://doi.org/10.1016/0034-4257(95)00186-7.

Rouse, W. J., Haas, R. H., Schell, J. A., Deering, D. W. (1974) Monitoring vegetation systems in the great plains with ERTS. Disponível em: https://ntrs.nasa.gov/search.jsp?R=19740022614

Schaefer, A.l., Dubeski, P.l.; Aalhus, J.l., Tong, A.k.w. (2001) Role of nutrition in reducing antemortem stress and meat quality aberrations. Journal Animal of Science. 79:91–101. https://doi.org/10.2527/jas2001.79E-SupplE91x

Schaefer, A., Genho, D., Clisdell, R.; Von Gaza, H., Desroches, G., Hiemer, L. (2018) The automated and real time use of infrared thermography in the detection and correction of DFD and fevers in cattle. Journal of Animal Science, 96(3), p.275–275. https://doi.org/10.1093/jas/sky404.604

Sivertsen, A. H., C. K.; Chu, L. C.; Wang, F.; Godtliebsen, K.; Heia, H.; Nilsen. (2009). Ridge detection with application to automaticfish fillet inspection. J. Food Eng. 90:317–324. https://doi.org/10.1016/j.jfoodeng.2008.06.035

Tong, A.k.w.; Schaefer, A.l.; Jones, S.d.m. (1995) Detection of poor quality beef with infrared thermography. Meat Focus Int. 4:443– 445.

Turner, T. A.; Eddy, L. (2001) Diagnóstico pela termografia. Revista Veterinária, n° 4, p. 17–95.

Valente J, Doldersum M, Roers C, Kooistra L (2019) Detecting rumex obtusifolius weed plants in grasslands from uav rgb imagery using deep learning. ISPRS Ann Photogramm Remote Sens Spat Inf Sci IV-2/W5:179–185. https://doi.org/10.5194/isprs-annals-IV-2-W5-179-2019

Webb EC, Erasmus LJ (2013) The effect of production system and management practices on the quality of meat products from ruminant livestock. S Afr J Anim Sci 43:413–423. https://doi.org/10.4314/sajas.v43i3.12

Wijesingha J, Astor T, Schulze-Brüningho D, Wengert M (2020) Predicting forage quality of grasslands using UAV-borne imaging spectroscopy. Remote Sens 12:126. https://doi.org/10.3390/rs12010126

Chapter 12
Internet of Things in Agriculture

José Augusto M. Nacif, Herlon Schmeiske de Oliveira, and Ricardo Ferreira

12.1 Introduction

The term "Internet of Things" (IoT) was first used by Kevin Ashton in 1999 and originally refers to reading Radio-Frequency IDentification (RFID) devices to track products. However, in 2015, the IEEE IoT working group defined the IoT as "an application domain integrating different technologic and social fields" (Minerva et al. 2015). Although generic, this definition refers to devices connected to the Internet that read and store environment data to automate tasks. More specifically, the IoT is an intelligent global network allowing: (1) interaction of devices capable of sensing, performing computation, and communicating with the Internet; (2) processing and information exchange between devices, data centers, and users; and (3) creating several intelligent systems (Zhang and Tao 2020). A CISCO report (Pepper 2019) estimates that by 2030 approximately 500 billion sensing devices will be connected to the Internet. These devices will be able to send data using IoT networks. The transferred data could be accounted for, evaluated, and distributed for processing using IoT services and applications. The IoT network supports different data formats and protocols and is always under evolution due to changes in people's daily lives. The main application scenarios of these devices include smart cities, transportation, agriculture, energy, industry, and health.

J. A. M. Nacif (✉)
UFV-Florestal Campus of the Universidade Federal de Viçosa, Florestal, Brazil
e-mail: jnacif@ufv.br

H. S. de Oliveira
Agrusdata, São Paulo, Brazil
e-mail: herlon@agrusdata.com

R. Ferreira
Informatics Department of the Universidade Federal de Viçosa, Viçosa, Brazil
e-mail: ricardo@ufv.br

© The Author(s), under exclusive license to Springer Nature
Switzerland AG 2022
D. Marçal de Queiroz et al. (eds.), *Digital Agriculture*,
https://doi.org/10.1007/978-3-031-14533-9_12

According to a McKinsey Institute study (Manyika et al. 2015), the economic impact of IoT systems should vary between 3.9 and 11.1 trillion US dollars (USD) until 2025. However, this value refers only to the transformation's economic impact and does not include IoT sales and services revenue. More specifically, the IoT global market for agriculture should reach 34.9 billion USD by 2027. Under the context of agriculture, it is possible to use IoT devices to facilitate the adoption of advanced technologies to make crops more sustainable and increase productivity by improving the efficiency of various stages of the process, such as optimizing the use of fertilizers. Digital agriculture uses a variety of IoT devices to collect, in real-time, the data generated during the crop development phase. The growth of the IoT solutions market for agriculture is tightly related to the recent adoption of precision techniques and the increasing use of embedded systems and cloud computing.

Several concepts are essential in the context of IoT systems. Ubiquitous or pervasive computing refers to the utilization of computing and information anytime, anywhere. Ubiquitous computing uses IoT devices to automate everyday tasks. For example, a ubiquitous computing system for home automation can integrate the lighting system with the air conditioning or heating equipment using environment or smart clothing sensors, thus continuously bringing thermal comfort to residents. Another example of ubiquitous computing is smart refrigerators that can read wireless RFID tags of products and suggest recipes based on available ingredients or compile shopping lists with consumed products. Invisible computing also relates to ubiquitous computing but refers to integrating computers into products in an imperceptible way.

Embedded systems are computers with reduced dimensions that are not visible to users. Their origin dates back to the 1970s and consists of hardware and software integrated into electrical or mechanical systems designed for a specific purpose. Embedded systems are primarily used in electronics such as TVs, cameras, smartphones, washing machines, and microwave ovens. Furthermore, embedded systems are employed in critical real-time applications performing different tasks, usually operating isolated, without integration with other real-time control functions. Cyber-physical systems integrate computing and physical processes (Guzman et al. 2020). Literature usually considers cyber-physical systems as an evolution of embedded systems but focusing on the physical world. It is important to note that although cyber-physical systems have communication, this capability is limited. On the other hand, systems based on the IoT include wireless communication, connection to the cloud, and the Internet. Figure 12.1 shows the relationship between these three terms.

Traditionally, embedded devices have limited processing capacity. As a result, many embedded systems often use the cloud computing paradigm, that is, they perform the computation on servers with large processing and storage capacity connected to the Internet. However, large-scale IoT systems generating vast volumes of data usually distribute the processing capability across the network, as illustrated in Fig. 12.2. Fog and edge computing are two alternatives to move the processing closer to the IoT devices that generate data and thus reducing the workload of cloud servers and the amount of data transmitted over the network. Fog computing

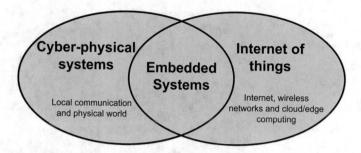

Fig. 12.1 Embedded systems, cyber-physical systems, and IoT

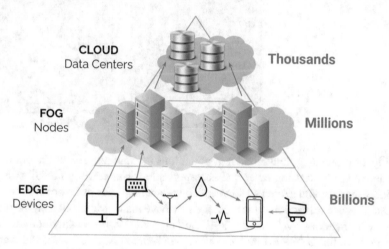

Fig. 12.2 IoT processing paradigms

performs data processing in nodes with higher computational power distributed across the network. In edge computing, the IoT device itself is responsible for processing the data it produces.

Figure 12.3a illustrates a three-tier architecture of an IoT system. The layers of this architecture are perception, network, and application (Mashal et al. 2015; Santos et al. 2016). The primary function of the perception layer, also known as the sensor or device layer, is to collect data such as location and temperature. The next level is the network layer responsible for providing a unique device address on the network and transmitting the data collected in the perception layer to each application and vice versa. This layer uses communication protocols such as Wi-Fi, Bluetooth, Zigbee, or others more focused on cloud computing. The last layer is the application layer that is responsible for providing services to clients. This layer manages IoT applications. Figure 12.3b presents an alternative five-tier architecture that includes the processing and business levels. This architecture adds the processing layer just above the network layer to store, analyze, and process the data produced by the devices. This layer uses several technologies such as databases,

Fig. 12.3 IoT
architectures: (**a**) three
layers; (**b**) five layers

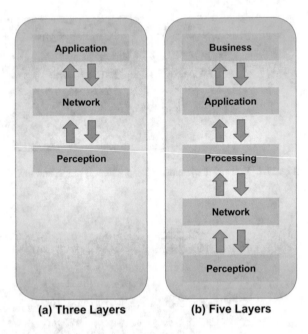

(a) Three Layers   (b) Five Layers

intelligent processing, and cloud and ubiquitous computing. The business layer is
just above the processing layer and is responsible for managing the IoT systems,
business models, and services and visualizing data in graphs or flowcharts.

Due to space restrictions in this chapter, we have made available an online repository with material available at https://github.com/lesc-ufv/iot4agriculture. This
repository contains detailed projects, including source code, software, and embedded systems technical documentation. In this chapter, we follow the international
standard IEC 80000-13 (IEC 2008), in which 1 kibibyte (KiB) equals 1024 bytes
(i.e., 2^{10}). IEC 80000-13 is fully compliant with other standards of the International
Bureau of Weights and Measures (BIPM) and the National Institute of Standards
and Technology (NIST).

12.2   IoT Devices

IoT devices and embedded systems belong to the perception layer. An embedded
system consists of four main components, as shown in Fig. 12.4a: processor, memory, input/output (I/O), and dedicated modules. The following sections detail each
one of the embedded system components. First, we focus on the perception layer
because this layer is fundamental to understand the demands of IoT systems for
specific applications, especially if we consider hundreds of embedded devices in a
heterogeneous scenario.

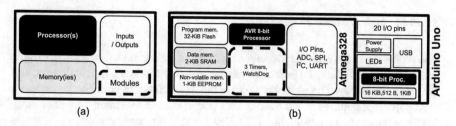

Fig. 12.4 Embedded system block diagram: (**a**) Four embedded system components; (**b**) Arduino UNO board

Figure 12.4b depicts a block diagram of one of the most popular embedded systems, the Arduino UNO, which contains four main components. The Arduino UNO has two microcontroller units (MCU). An MCU is a tiny embedded system including a processor core, memories, I/O, and modules in a single integrated circuit package. The main MCU of the Arduino board is an Atmega328 processor with 32 KiB of program memory, 2 KiB of data memory, and 1 KiB of non-volatile memory, in addition to I/O modules such as ADC, serial protocols, and timers. The second MCU is an ATmega16U2 processor with 16 KiB of program memory, 1 KiB of data memory, and 512 bytes of non-volatile memory. This MCU has the specific task of providing the programming interface for the main MCU and the USB port. The UNO board also offers power supply circuits, LEDs, and I/O pins to connect external devices.

There are some important details of the Arduino term (Barrett 2020). Nowadays, it is common to use Arduino for a low-cost embedded system board. Nevertheless, the Arduino proposal goes beyond a low-cost board. From the 1980s until 2005, embedded system programming and validation was a difficult task. The Arduino project started in 2005 in Ivrea, Italy, by introducing new ideas to simplify embedded systems development. First, the project includes hardware and software as an open-source platform. Arduino initially was based on a simple extension of C/C++ languages. Second, Arduino creates standardization for all sensor and actuator libraries, reduces development costs, and delivers an entirely integrated system design. Most sensors are now compatible with standard libraries and offer code examples. When the programmer installs a new sensor library, there are code examples available to connect and read/write data to/from the sensor. The central idea is learning by example, and then the programmer modifies the primary example to adapt to new application requirements. Third, using a standard library increases the design portability across diverse embedded system boards and MCUs. For instance, it is possible to use the same C language code for both an Arduino UNO and ESP32 processor without any modification (Beroleti 2019). These abilities contribute significatively to the wide use of Arduino-based systems and tools. Finally, a single environment provides resources to write and deploy the program, and at the same time, validate the embedded system.

12.2.1   IoT Device Processor

The processor is the most important component of an embedded system. The most relevant characteristics of a processor are the word bit width, the instruction set, and the clock frequency. The word bit width is the basic compute unit. For instance, an 8-bit processor requires at least one clock cycle to compute an integer number ranging from 0 to 255. Therefore, any operation with 32 bits numbers, such as a single float number, requires several clock cycles to be computed. The clock cycle is the time unit, which is the inverse of the clock frequency. For instance, a clock frequency of 16 MHz has a clock cycle of 63 ns. Therefore, even a low-cost, low-frequency, 8-bit processor such as an Arduino UNO running at 16 MHz is fast enough to process data for most embedded system applications.

Furthermore, low clock frequencies reduce the power consumption and the heating, increasing autonomy without battery recharging or power supply (Zuquim et al. 2003; Caldas et al. 2005). Finally, the instruction set or instruction set architecture (ISA) defines the interface between the software and hardware to ensure binary compatibility. The most famous ISA example is the Intel ×86, which has kept all binary executable programs compatible with any ×86 successor since the 1980s. Nevertheless, in general, embedded processors do not require an operating system layer such as Linux or Windows. Therefore, compiling to a new ISA is almost straightforward using the source code in C language.

Currently, low-cost MCUs with 32-bit processors have become popular. The main reason is the complexity of the Wi-Fi protocol (IEEE 802.15), which requires high-speed processors, such as ESP8286 or NodeMCU, but that cost less than two USD. Furthermore, these platforms usually have multiple processors or cores inside an MCU. For instance, the ESP32, which is an evolution of the ESP8266, has three processors. One of them manages the Wi-Fi communication, and therefore the main processor is exclusively responsible for executing the user tasks. In addition, a third simple and low-power processor is available to switch off the two 32-bit main processors to operate in waiting or sleeping mode to save energy.

12.2.2   IoT Device Memories

There are four main types of memory used in embedded systems: program memory, data memory, non-volatile memory, and external memory. Non-volatile memory keeps the data without requiring any power supply. However, this type of memory is slow and has a maximum number of write cycles. In general, the non-volatile memory stores parameters, crucial data, passwords, and initial settings. The program memory stores the user code, and it is written once at compile time. It is also possible to store constant data in program memory to save space in data memory. These memories usually are read-only memory (ROM). The data memory uses random access memory (RAM) and usually has a reduced size in embedded systems to reduce the overall costs. Finally, external memory offers support to more complex applications. For example, it is possible to use an SD card to expand the data memory capacity significantly.

Most program memories are non-volatile to store the program binary code. The first MCU generations use PROM memories (Programmable Read-Only Memory) or EPROM (Erasable PROM), which flash memories have replaced. The flash memories increase the capacity significantly, are non-volatile, erasable, and do not require complex circuits to write new data. Moreover, flash is a low-cost memory since it has become popular with SD cards and memory sticks.

One of the main challenges in the first MCU generation was the code upload to the program memory. In general, a specific external board was required, as shown in Fig. 12.5a. During the last two decades, MCU companies have significantly simplified this step. For instance, Arduino incorporates the programming module inside the embedded system board. In a few seconds, the user uploads the binary code using a simple USB cable to connect the computer and the embedded system board, as shown in Fig. 12.5a. A third option is the Over the Air (OTA) programming mode using a wireless connection (for instance, Wi-Fi) to update the program binary code.

The data memory stores the values at runtime, and most MCUs have low-capacity memories, ranging from a few bytes to 64 KiB. These memories use Static Random Access Memory (SRAM) technology to provide a clock frequency compatible with the processor. However, these memories are volatile, and they require a constant power supply to avoid data loss.

The non-volatile memories store the permanent data using Electrically Erasable Programmable Read-Only Memory (EEPROM) technologies. Most MCUs have from 512 to 2 KiB EEPROM memory. These devices have low clock frequency and a maximum number of writes. Figure 12.5b shows a typical embedded memory system.

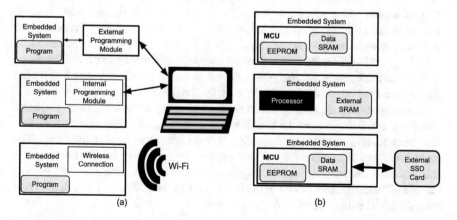

Fig. 12.5 Embedded systems architecture and programming: (**a**) Three programming modes; (**b**) Three examples of memory module organization

12.2.3   IoT Input/Output

The input/output (I/O) signals provide the interaction between the embedded system and external devices. An I/O interface consists of one or more digital or analog signals. The data communication can be synchronous or asynchronous, serial or parallel, wired or wireless. In addition, an I/O signal can implement several functionalities dynamically selected at runtime. The MCU processor or other independent hardware module can implement the interface protocols.

An input analog signal ranges from 0 to V_{max} Volts. The Analog to Digital Converter (ADC) converts this signal to a digital value. The two most important ADC features are the resolution and sampling rate. The resolution defines how to transform a continuous analog into a discrete value. For instance, a resolution of 10 bits (or $2^{10} = 1024$ values) for a range of 5 V results in a resolution step value of 5 V/1024 = 4.88 mV. The sampling rate defines the maximal number of input samples per time unit. For example, Arduino Uno has six 10-bit converters with a sampling rate of 10,000 samples per second. An analog output signal has a Digital to Analog Converter (DAC) to generate an analog value ranging from 0 to V_{max} by converting a discrete digital n-bit value. A DAC also has a resolution of n-bits and a sample rate. However, fewer embedded systems have DAC modules at the same MCU package compared to ADC modules. Temperature sensors such as thermistors and light sensors (LDRs – Light Dependent Resistors) are two common analog sensor examples that require an ADC module interface.

It is possible to use a single-bit digital I/O interface to connect, for example, a switch button (input) or an LED (output) to perform simple input/output tasks. Also, a digital signal can generate a pulse time sequence with a width modulation known as Pulse Width Modulation (PWM). This interface is useful for controlling motor speeds or light brightness by using a single-bit digital signal. A PWM signal also emulates an analog value by using the average wave voltage. For instance, in an Arduino UNO, the command WriteAnalog generates a PWM output signal to emulate an analog value since this MCU does not have a DAC hardware module. Furthermore, a digital signal also measures time intervals or counts the number of pulses to measure a motor speed. Finally, it is important to highlight that an MCU signal supports maximum current values that should not exceed a few mA. Therefore, you should incorporate external power drivers in your embedded system if you need to use high-load power devices. In general, the maximum voltage is 3.3 V or 5 V.

Furthermore, any digital signal can handle a bit sequence as serial data. The communication protocol defines the base frequency, configuration bits, data packets format, control signals, and communication type (synchronous or asynchronous). The main advantage of a serial protocol is to use fewer wires. However, the transmission rate is usually not feasible for some applications needing higher bandwidth, such as cameras. Therefore, some low-cost cameras, such as the ESP-CAM board, interface with the MCU using a parallel 8-bit interface.

12.2.4 Specific Hardware Unit in IoT Devices

Embedded systems have dedicated hardware modules to perform specific tasks. The main goal is to help the primary MCU processor by implementing functionalities directly in hardware. The most common module is the timer. For instance, the Arduino UNO has three timers. In addition, a timer is frequently employed in the perception layer to control measurement intervals. With the evolution of embedded systems, as a result, more resources have been added, including hardware units to perform serial protocols such as UART (Universal Asynchronous Receiver/ Transmitter), I²C (Inter-Integrated Circuit), and SPI (Serial Peripheral Interface). Recently, MCUs such as K210 include hardware units for cryptography or Fast Fourier Transform (FFT). These hardware units execute tasks in parallel to the MCU processors to satisfy real-time constraints.

Embedded systems offer hardware support for many standard communication protocols. The most usual modules are I²C, SPI, CAN (Controller Area Network), One-Wire, and UART. While SPI, I²C, and One-wire allow multiple point connections, the UART protocol is point-to-point. Figure 12.6a shows an advantage of multiple point connections (many-to-many). The most common applications for point-to-point connections are to upload the MCU code and provide wireless communication. Figure 12.6b depicts a point-to-point connection between two devices, A and B.

Table 12.1 shows the main protocol features. The SPI protocol has an additional control signal for each slave device. A synchronous protocol has a clock signal, and the asynchronous protocol does not. In general, a protocol can have one master node and multiple slave nodes. However, it is also possible to switch the master control node in some protocols. The network covers areas ranging from a few meters (short) to dozens of meters (long), which depends on the cable properties (capacitance, for instance). The last column presents devices that usually employ these protocols.

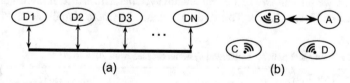

Fig. 12.6 Device connection types: (**a**) Multiple-points or many-to-many connection; (**b**) Point-to-Point connection.

Table 12.1 Main serial protocol features

	Signals	Clock	Master/Slave	Cover area	Devices
UART	2	No	Point-to-point	Short	Radio, programming
I²C	2	Yes	Many-to-many	Long	Sensors, displays
SPI	4+	Yes	One-to-many	Short	SSD card, displays, memories
1-Wire	1	No	One-to-many	Long	Sensors
CAN	2	No	Many-to-many	Medium	Sensors

12.2.5 IoT Device Examples

This section presents the most popular embedded systems from 2010 to 2020: Arduino Uno, ESP8266/NodeMCU, ESP32, and Raspberry Pi. In addition, we also introduce two new promising boards which offer support to artificial intelligence: Arduino Nano 33 IoT and SoC K210.

Arduino UNO is a low-cost board, by far the most popular embedded system in the decade. Although it is a highly resource-constrained MCU (8-bit processor and 2 KiB of data memory) and does not support wireless connection, the Arduino UNO simplicity, many examples, and tutorials are key features for reaching this level of popularity. The ESP8266, or NodeMCU, was the first low-cost MCU to support Wi-Fi. The ESP32 is the second generation of the ESP8266 by improving the primary processor performance, memory size, integrated sensors (touch and temperature sensors), DAC, and Bluetooth. Finally, the Raspberry PI is a low-cost and high-performance embedded system that can execute operating systems (Linux and Windows), supporting video output, external cameras, and SSD cards.

Table 12.2 shows the main characteristics of the selected embedded systems. ESP32 and ESP8266 have good cost-benefit ratios. The programming memory supports a simple data file structure, and it is possible to update the software using a wireless connection. Both boards support C/C++ and MicroPython languages. While the Arduino Uno does not offer Wi-Fi, it requires an external module for Wi-Fi, and the most usual option if Wi-Fi is needed is to add an ESP8266 board. The main advantage of an Uno board is the robustness, powered by 5 V to directly connect sensors and actuators, while the ESP32, ESP8266, and Raspberry interfaces only support 3.3 V. A Raspberry Pi is a complete computer inside an embedded system and an attractive solution for an IoT computer server. There are more than 200 options for similar embedded systems. Figure 12.7 shows the four selected boards in scale.

Finally, we present two recent embedded systems with artificial intelligence features: Nano 33 and K210. The Arduino Nano 33 (more details at https://store.

Table 12.2 Four most popular embedded systems devices from 2010 to 2020

Sistema	Cost (USD)	Processor	I/O and modules	Wireless	Clock	RAM	Prog. mem
Arduino Uno	3	ATmega 324 (8 bits)	UART, SPI, I²C, 6 ADCs	Não	16 MHz	2 KiB	32 KiB
ESP8266	2	L106 (32 bits)	UART, SPI, I²C, 1 ADC	Wi-Fi 802.11	80 MHz	160 KiB	4 MiB
ESP32	3	LX6 (32 bits, 2 cores)	UART, SPI, I²C, 18 ADCs, 2 DACs, *Touch*, Temperature	Wi-Fi 802.11 BLE 4.2	160 MHz	512 KiB	4 MiB
Raspberry Pi 3B	30	ARM A53 (64 bits, 4 cores)	UART, HDMI, camera	Wi-Fi 802.11 BLE 4.2	1,4 GHz	1 GiB	SSD

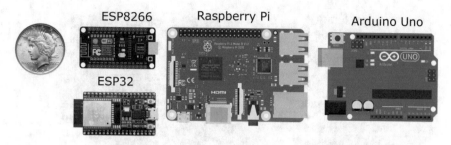

Fig. 12.7 Four popular embedded system boards: Arduino Uno, ESP8266 (or NodeMCU), ESP32, and Raspberry Pi3 B (The coin diameter is 30.6 mm)

arduino.cc/usa/nano-33-iot) has three processors. The central processor has a core ARM Cortex-M0 32 bits, 48 MHz, 256 KiB of programming memory, and 32 KiB data memory. There is also a NINA-W10 processor with two 32-bit cores to manage the wireless Wi-Fi 802.11b/g/n and Bluetooth 4.2 communications, a co-processor for cryptography, and a hardware unit with a 3D accelerometer and gyroscope. However, the Nano 33 cost is significantly higher than an ESP32, around five to six times more. In addition, the Nano 33 supports the Google Tensorflow lite APIs. The second board uses K210 MCU, which currently costs from 15 to 30 USD. However, it offers a dedicated input port for the camera and a hardware unit to perform 1×1 and 3×3 convolution for deep learning computation. The K210 performance is 5000–60,000 times faster than the four embedded systems depicted in Table 12.2, except the Raspberry Pi 3B, which is 500–1000 slower than the K210 to perform neural network computations. Moreover, the K210 is a low-power processor (0.3 W) in comparison to a Raspberry PI that consumes three to six times more power (1–2 W).

Figure 12.8 shows two examples of embedded systems based on the K210 MCU that include a graphical color display, standard I/O connectors, buttons, and battery, allowing cost ranges from 20 to 40 USD. The m5stickV device has 200 mAh Lipo Battery, 1.14 TFT, 135×240 color display, and OV7740 camera. The Maix Amigo device looks like a simple smartphone with a graphical color touch 320×240 display.

12.3   Communication and Connectivity in IoT Systems

An IoT system can make use of both wired and wireless connections. Depending on the scenario, wired connections can be easily installed and significantly reduce energy consumption. There are several wired connection standards that both embedded systems and commercially available sensors and actuators support, as described in Sect. 12.2.4. On the other hand, wireless connections offer a more straightforward installation process. However, in precision agriculture and digital agriculture, there are usually many obstacles such as vegetation and natural barriers that can

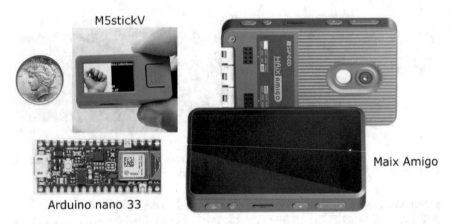

Fig. 12.8 Arduino Nano 33 IoT and two embedded systems based on K210 MCU: m5stickV and Maix Amigo

degrade the quality of communication. Another aspect is energy consumption since battery-power devices appear often. Some embedded systems already include the communication interface, and others need additional modules. The main considerations when comparing are cost, range, mode of operation, and energy consumption.

12.3.1 IoT Short-Range Communication

Bluetooth, Wi-Fi, and infrared protocols are popular short-range IoT protocols. Wi-Fi implementation follows the 802.11 protocol capable of reaching 300 m but is generally limited to less than 100 m. In addition, ESP8266 and ESP32 low-cost systems further simplify the creation of Wi-Fi networks in an IoT environment. Another advantage is easily connecting to devices such as smartphones or personal digital assistants. However, on the other hand, Wi-Fi power consumption is prohibitively high for many IoT scenarios.

Bluetooth is another widespread protocol used in the IoT domain, offering good communication performance combined with low power consumption (Alves et al. 2019). For example, an ESP8266 device with Wi-Fi consumes between 100 and 200 mW, while Bluetooth needs around 1–2 mW, for a range of 1–10 m. Using Bluetooth instead of Wi-Fi can result in up to 100 times power consumption reduction in various scenarios. The Bluetooth range can reach 100 m, although increasing the power demand to 100 mW.

It is also worth mentioning that some embedded devices offer mesh protocols. For example, ESP32 and ESP8266 support ESP-Wi-Fi-Mesh and can connect up to 1000 nodes in an area without any additional Wi-Fi support. Another option is ESP-NOW, which allows the connection of multiple ESP devices without a Wi-Fi connection. Similar to remote controls, IoT devices also use infrared LEDs/sensors for communication. However, despite the very low power consumption, the range is limited, and communication requires aligned devices with no obstacles. Recently, Visual Light Communication has emerged as an alternative option to infrared communication (Matheus et al. 2019).

There are several other options for low-power, short-range communication. An example is the NRF24L01, with applications such as drone and IoT communication protocol. This low-cost radio (~1 USD) has an SPI interface to connect most embedded systems and a transmission rate of up to 2 Mbps with a range of up to 100 m and the possibility of creating networks with hundreds of nodes.

Another option for short-distance applications is the RFID that performs radio frequency identification. The system consists of two types of modules: RFID tags and RFID base. It is possible to place an RFID tag on an animal or product or even on a card for identification. The tag works passively and responds to the signal sent by the base. In general, the RFID tag does not need a battery, although semi-passive or battery-active tags also exist.

12.3.2 IoT Long-Range Communication

Zigbee network (XBee), LoRa (LOng RAnge), Wi-Fi interface, and satellite are the most used options for long-range communication. Wi-Fi is an alternative to building sensor networks with distances of kilometers. However, this solution is highly power-hungry, which is the most significant challenge beyond the cost of 50 USD for each Wi-Fi unit. In this section, we introduce the two most popular solutions: Zigbee and LoRa.

The Zigbee network supports connecting multiple devices from tens of meters to a few kilometers apart. The advantage over a Bluetooth network is that Zigbee is at least 10 times more effective in power consumption. With 1 mW, we can transmit to a range of 100 m, while Bluetooth is limited to 1–10 m. However, Zigbee is highly more costly than Bluetooth, ranging from 12 to 15 USD, for a radio module already available in an ESP32 at the cost of 2–3 USD. The Zigbee radio works in a range from 2 to 4 km, consuming around 80 mW.

The LoRa WAN network adopts LoRa technology which uses spread spectrum modulation. Military and space applications previously used this technology, and recently, several companies have adapted it for low-cost commercial use. LoRa allows transmissions of more than 10 km in rural areas with low power consumption. Another advantage of LoRA is the availability of embedded systems with LoRa included (Beroleti 2019). In addition, these cards support Wi-Fi and Bluetooth, bridging a local area network with a wide area network into a low-cost system in the 12 USD range. Other alternatives are GSM modules or smartphones to connect the IoT devices using Bluetooth or Wi-Fi. In the next decade, 5G will play an essential role in digital agriculture, improving yields and crop quality and, at the same time, minimizing labor necessity (Tang et al. 2021). 5G is faster, more stable, and security-improved to deliver new products and services. Furthermore, 5G provides better infrastructure for self-driving vehicles and AI devices. According to the GSMA (2021), 5G connections will be 20% of global commercial connections by 2025.

12.3.3   IoT Cloud Connection

In addition to devices and their communication, an IoT system includes an Internet connection. In this section, we introduce the main protocols and services for cloud integration.

12.3.3.1   IoT Cloud Protocols

In addition to Internet protocols, such as TCP/IP and UDP/IP, lighter protocols with functionalities for sensor networks have also been developed. Among the protocols for IoT, we highlight the Message Queuing Telemetry Transport (MQTT), an open standard ISO/IEC 20922. MQTT is a lightweight protocol based on the exchange of messages between devices and uses publication and subscription (publish-subscribe) mechanisms.

We will use an example to explain the MQTT protocol. Suppose an experiment with two greenhouses equipped with temperature and relative humidity sensors, a fan, and a humidifier. We can use the DHT11 sensor to measure relative humidity and air temperature. Suppose the fan has three operating speeds and the humidifier has only two modes: on and off. An MCU is available in each greenhouse. Suppose that greenhouse one uses a NodeMCU module and that greenhouse two counts with an ESP32 MCU. Figure 12.9 illustrates this scenario.

First, the MQTT server (broker) must be available on a computer. This server can be a desktop computer, as illustrated in Fig. 12.9. There are other more cost-effective options in which the MQTT server can be installed on a low-cost embedded platform such as a Raspberry PI board or can even use a public MQTT server in the cloud. The most used MQTT server is the Mosquitto which runs on Windows, Linux, or macOS operating systems. The server's role is to receive and forward messages to clients. Mosquitto supports multiple clients, and there is no limit. Figure 12.9 shows an example with two greenhouses, a smartphone, and a notebook. To communicate, the customer can send and receive messages. Each message has an associated topic and can have a hierarchical structure similar to organizing files on a computer with folders and subfolders.

The system follows a subscription and publication scheme. First, the customer subscribes to and receives from a thread. For example, a mobile customer can subscribe to the topic "/greenhouse/1/indoor/temperature" to receive information about the internal temperature of greenhouse 1. The greenhouse 1 posts messages on the topic "/greenhouse/1/indoor/temperature" with data "32" informing that the

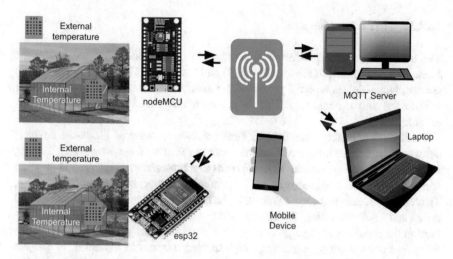

Fig. 12.9 Example of an IoT scenario with two embedded systems, MQTT protocol, and connection to smartphones and laptops

temperature is 32 °C in greenhouse 1. Every time greenhouse 1 publishes the temperature, all customers subscribing to this topic receive the message. For example, the laptop might also be subscribing to this topic or a more general topic such as "/ greenhouse/1." Every time greenhouse 1 publishes temperature or relative humidity information in a sub-topic, the customer who subscribed to the general topic receives the message.

MQTT has three significant advantages. First, MQTT is a scalable solution where you can easily add or remove clients or sensors. By definition, MQTT is a many-to-many communication protocol, which is ideal for IoT applications. Second, MQTT supports the hierarchical organization with topics and subtopics. Thirdly, there are many supporting applications and frameworks from both the sensor node and the client node sides. For example, to design a client application, the developer can choose from several user-friendly interfaces to create dashboards for Android and iOS mobile platforms. Most environments with cloud dashboards for IoT support MQTT. From the sensor node point of view, MQTT libraries have a high-level interface with the main functionalities: (1) connect to an MQTT server (broker), (2) subscribe to a topic (subscribe), and (3) publish data in a topic (publish). Finally, several applications develop a graphical visual interface for using MQTT in Android, iOS, or mobile operating systems.

The Constrained Application Protocol (CoAP) is a widely used IoT protocol simplifying Web integration for multiple destinations using the HTTP protocol. CoAP runs on most devices that support the UDP protocol. While CoAP is a one-to-one protocol in the client-server format, MQTT is many-to-many, in which messages pass through the broker. MQTT messages are all-purpose, but clients need to know the formats in advance. Both protocols have benefits and disadvantages depending on the application.

12.3.3.2 IoT Cloud Services

This section introduces platforms supporting IoT in the cloud. First, we present the Node-Red environment (Passe et al. 2017) and then the IFTTT (If This Then That) service. Node-Red is an IBM project that visually integrates various resources for IoT design and execution. IFTTT is a methodology that supports event-based programming.

Node-Red connects visually and straightforwardly several hardware devices; software (database, e-mail); and online services (weather forecast website, Twitter, Facebook, Telegram). The environment runs in a browser allowing project creation and node connection. A node can be, for example, a temperature sensor, a database, Twitter, or a weather forecast website. The Node-Red library provides over 3000 nodes and 1800 examples. To create small actions, the programmer can use functions in the JavaScript language or other languages such as Python. Node-Red intuitively supports saving, executing, and reusing nodes. In addition to project visualization, Node-Red can add dashboards supported by browsers, thus allowing desktops and smartphones to interact with IoT projects.

Figure 12.10a presents a Node-Red project example with a temperature sensor sending values using an MQTT connection (sensor node). The system stores the temperature values in a MongoDB database and allows visualization using a control panel (Show Temp node). A programming node checks if the temperature is less than 15 °C. If positive, it notifies another node to generate an e-mail message and turn on the heater, which also connects to the control panel allowing manual operation. Finally, the system collects and displays in the control panel weather data from a forecasting website. The weather forecast also connects to the node that triggers the heater, sending messages to notify the user by e-mail. This example illustrates the simplicity of the interface for project development, execution, and maintenance. Figure 12.10b shows an example control panel displaying graphics, dials, icons, and control devices with a slider switch.

IFTTT is a service with several applications in IoT integration and allows programming actions to respond to various types of events over the Internet. A simple

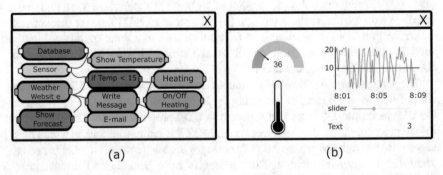

(a)   (b)

Fig. 12.10 Node-Red development environment: (**a**) Node-Red project example; (**b**) Control panel example

example is to create an event to monitor a weather forecast website and send an e-mail if there is a rain forecast for tomorrow. Through IFTTT, you can connect to several services such as Amazon's Alexa, Google Assistant, Twitter, Dropbox, or Google calendar. IFTTT is widespread and has many examples of projects available on the Internet. Node-Red and IFTTT are open platforms and services. However, there are several commercial platforms available.

ThingSpeak started as a free platform until its acquisition by Matlab. This service aims to provide a simple interface to visualize, analyze, and store data. It is also possible to control devices with the MQTT protocol. Another cloud service with a friendly and straightforward interface is Ubidots, which supports several embedded systems with code generation and tools to manage events; store data intuitively; and send e-mail, SMS, or Telegram messages. Specifically designed for smartphones, the Blynk application provides an infrastructure for IoT. The application supports various embedded systems with different sensors/actuators, cloud storage, and a user-friendly interface.

Large information technology companies such as Google, Amazon, and Microsoft also offer a wide range of services. For example, Google Cloud IoT provides a set of tools to process, store, and analyze data, including Google services (e-mail, spreadsheets, artificial intelligence for recognizing images and sounds). Also, Microsoft offers a wide range of features in the Azure platform for IoT, and Amazon Web Services provides general services integrated with specific services for IoT.

12.4 IoT Applications in Agriculture

IoT systems can improve several farming activities in agriculture, commonly following the pattern illustrated in Fig. 12.11. The general scenario includes several IoT sensors sending data to servers in the cloud. Then, data is processed, thus providing decision-making information through applications or other computers. The main applications can be divided into categories (Kim et al. 2020): (a) management systems, (b) monitoring systems, (c) control systems, and (d) unmanned Machines.

12.4.1 Management Systems

IoT devices provide real-time data to manage cultivation, fleets of agricultural machinery, and energy and water consumption more efficiently. IoT-based agricultural information management systems help farmers make decisions regarding cultivation. IoT devices generate data about machinery, seeds, pesticides, and fertilizers. The most used types of sensors are for measuring soil moisture, temperature, and pH. Subsequently, the farmers can perform financial analyses using big data and machine-learning techniques to optimize profitability. Also, it is possible to use IoT

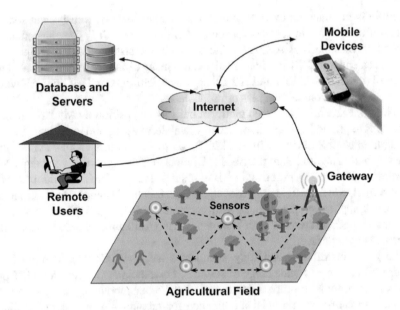

Fig. 12.11 General scenario of IoT applications in agriculture

to manage water consumption and reservoir levels, reducing water demand by up to 60%. This application uses soil moisture, temperature, flow, and water level (ultrasonic) sensors. Furthermore, GPS sensors in vehicles can improve agricultural machinery management by optimizing the path and operation of these machines, reducing costs, and increasing productivity.

Figure 12.12 shows an example of the FUSE commercial system developed by AGCO company that uses IoT to automate the management of the agricultural machinery fleet (more information at https://www.fusesmartfarming.com). This system offers software platforms to manage agricultural machinery in the various stages of the production process: soil preparation, planting, crop development, harvesting, and storage. Each machine has a GPS and a communication channel that allows real-time transmission of vehicle parameters, such as speed, location, consumption, and available fuel estimation. This system reduces production costs by up to 22%.

12.4.2 Monitoring Systems

The main applications of monitoring systems are mostly related to diseases, fields, greenhouses, cattle, pests, and soils (Gull et al. 2016). Several systems use IoT devices to predict disease in crops. Field monitoring can be useful to improve quality and to obtain productivity maps to optimize farm profit. The most common sensors used in field monitoring are humidity (environment and soil), temperature,

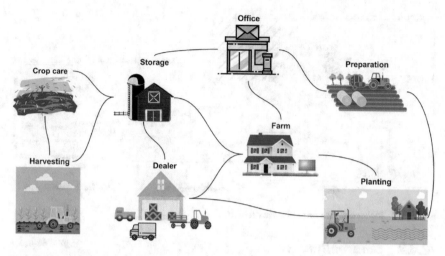

Fig. 12.12 Agricultural machinery management system

CO_2, light intensity, and cameras. In greenhouses, environmental conditions of temperature and humidity affect plant quality and productivity. In addition to temperature and humidity sensors, monitoring systems use light intensity, pressure, and CO_2 sensors. IoT monitoring systems are also useful to collect data generated by livestock and poultry herds. For example, there are systems for monitoring both the fertile period and the time cows are calving based on tracking the movement pattern of the animals. For birds, a typical monitoring system includes temperature, air velocity, ventilation rate, humidity, and concentration of gases such as carbon dioxide and ammonia.

The main application of IoT pest monitoring systems is the early prediction of significant infestation events, thus optimizing pesticide use. The primary sensors in these systems are humidity, lighting, temperature, hyperspectral, acoustic, and infrared cameras. Soil quality has a direct impact on plant growth and productivity. Thus, soil monitoring with IoT devices is essential to improve planting efficiency and agricultural management. Examples of soil monitoring systems use pH, temperature, and humidity sensors to improve irrigation control decisions and pesticide application. Another critical optimization factor for soil monitoring is defining the timing and amount of fertilizer needed in different scenarios. In addition to those already mentioned, these systems can use ultraviolet, ultrasonic, and nitrogen-phosphorus-potassium (N-P-K) sensors.

Figure 12.13 presents a system for predicting plant diseases based on IoT devices proposed by Khattab et al. (2019). The main parameters measured by the sensors are air temperature and humidity, soil temperature and humidity, wind speed and direction, amount of rain, solar radiation, and leaf moisture. The sensor transmits data using GPRS/GSM/3G/4G wireless technologies to be later used in training machine learning models to predict diseases.

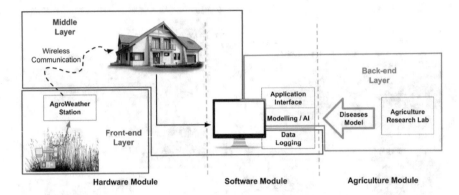

Fig. 12.13 IoT system for crop disease prediction

12.4.3 Control Systems

IoT-based control systems are useful to automatically manage crop fields, greenhouses, irrigation, and water quality, thus maintaining optimal crop growth (Vieira et al. 2018). Crop fields use sensors for gas, movement, temperature, humidity, water flow, level, and soil fertility. Most of these sensors are autonomous and can control actuators. Controlling the greenhouses' environment and generating productivity maps result in crop quality and profit improvement. Some systems automatically control temperature and humidity so that the greenhouse is always in ideal working conditions. The primary sensors used for monitoring greenhouses are luminosity, temperature, humidity, conductivity, and CO_2. Irrigation control systems utilize IoT devices to manage water resources efficiently. These systems monitor soil moisture and other external factors, such as weather forecasting, to irrigate crops without wasting water and, at the same time, not reducing productivity due to water stress. Water quality control systems use pH sensors to reuse water in agricultural systems. The main sensors of this type of system are pH and temperature.

Figure 12.14 presents an example of a greenhouse control system for orchids (Liao et al. 2017). The system's objective is to find and maintain the ideal growing conditions for orchids by employing cameras and light, temperature, and humidity sensors connected to the cloud using the Wi-Fi and Zigbee protocols. The results demonstrate that the ideal growing conditions are the temperature of 28.8 °C and 71.8% of relative humidity, allowing an average leaf growth rate of 79.4 mm^2 per day.

12.4.4 Autonomous Agricultural Machinery

Autonomous agricultural vehicles have been developed since the 1980s and use various sensing systems. Several companies have invested in GPS-guided autonomous tractors. The advantages of autonomous tractors include pre-planning,

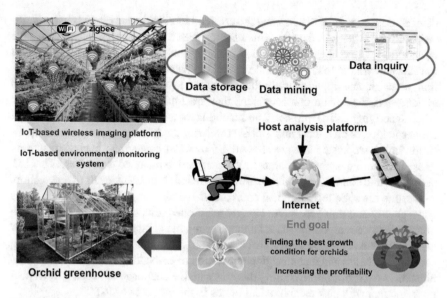

Fig. 12.14 Greenhouse IoT control system orchid growing

optimizing the path, and improving the quality of operation in low visibility conditions. With recent advances in IoT systems, the development of autonomous tractors has reached a new level, allowing multiple machines to be connected and exchange data. For example, various tractors can be combined and communicate to mimic the route and speed of the leading tractor for simultaneous operation. Unmanned aerial vehicles or drones utilize IoT systems and have contributed to taking agriculture to new levels of intelligence in precision farming and digital farming. These vehicles have been widely used for environmental monitoring, identifying pest and disease outbreaks using spectral images of crop fields. In this direction, drones can also take advantage of thermal cameras to provide data on wildlife, plants, diseases, and water.

For example, the John Deere company has a series of integrated systems to help automate the tasks performed by tractors. The Machine Sync system allows tractors to communicate directly and, combined with other systems, can improve harvesting efficiency and accuracy. The AutoTrac Vision system includes the functionality of tractors to follow planting lines, reducing potential damage and improving efficiency. Finally, the AutoTrac RowSense system avoids hitting plants and ensures complete fertilizer coverage.

12.5   Final Remarks

The emergence of the IoT has created new possibilities for research and applications in precision and digital agriculture with the reading, storage, and integration of data at local and global scales. The processing and exchange of information between devices, data centers, and users and the artificial intelligence revolution will significantly impact the next decade. This chapter presents the subject by showing the intersection with embedded systems, cyber-physical systems, and the new concept introduced with the IoT. We have structured the text in a three-layer model: perception, network, and application. In the perception layer, we highlight the embedded systems and their main characteristics that are fundamental for decision-making when specifying an IoT project. The second layer addresses the network layer that connects devices on a local and global scale with the Internet. Finally, in the application layer, we systematically show the four main categories: management systems, monitoring systems, control systems, and unmanned machines. The IoT permeates and connects various areas of knowledge at different levels of abstraction.

Despite the solid technological component, we seek to structure the IoT models conceptually and universally. As already mentioned, this chapter also has an additional reference with online material that is available at https://github.com/lesc-ufv/iot4agriculture. This repository provides detailed examples of projects, including code and software and technical documentation of embedded systems and communication networks, complementing this text with the challenge of keeping the repository updated with new technological trends in the vast area of IoT.

Abbreviations/Definitions

- Telemetry: is the remote collection of data obtained by sensors or some other source and its transmission to some equipment that receives and makes use of that data.
- CAN bus: is a system that creates communication pathways between the electronic control units within a vehicle, allowing the transfer and interpretation of collected data.
- Powertrain: is an assembly of every component that pushes your vehicle forward. Your car's powertrain creates power from the engine and delivers it to the wheels on the ground. The key components of a powertrain include an engine, transmission, driveshaft, axles, and differential.
- GPRS: is a packet-oriented mobile data standard on the 2G and 3G cellular communication network's global system for mobile communications (GSM).
- 4G: is the fourth generation of broadband cellular network technology, succeeding 3G and preceding 5G. First released to the public in 2009, 4G networks offered significantly faster data rates, lower latency, and more efficient use of the radio frequency spectrum.
- 5G: is the fifth generation of broadband cellular network technology, succeeding 4G. 5G enables a new kind of network that is designed to connect virtually everyone and everything together including machines, objects, and devices. 5G wireless technology is meant to deliver higher multi-Gbps peak data speeds, ultra-low latency, more reliability, massive network capacity, increased availability, and a more uniform user experience to more users.
- Cloud computing: is the on-demand availability of computer system resources, especially data storage (cloud storage) and computing power, without direct active management by the user.

- Value chain: the full range of activities that firms and workers do to bring a product from its conception to its end use and beyond. This includes activities such as design, production, marketing, distribution, and support to the final consumer.
- Standardization: is the process of developing and implementing technical standards. Standards increase compatibility and interoperability between products, allowing information to be shared within a larger network and attracting more consumers to use the new technology, enhancing network effects.
- Digitalization: refers to taking analog information and encoding it into zeroes and ones so that computers can store, process, and transmit such information.
- Self-propelled: equipment that has its own power source to move or operate. Common examples are harvesters (combines) and self-propelled sprayers.
- Decision matrix: is a list of values in rows and columns that allows an analyst to systematically identify, analyze, and rate the performance of relationships between sets of values and information. Elements of a decision matrix show decisions based on certain decision criteria. The matrix is useful for looking at large masses of decision factors and assessing each factor's relative significance.
- Life cycle: the phases of development through which a computer-based system passes.
- Digital Agriculture: It´s all the digital tools applied to collect, store, analyze, and share electronic data and/or information to increase agriculture productivity and help the decision-making.
- Small Data: is data that is "small" enough that can be understood by humans, or in other words, it is data in a volume and format that makes it accessible to extract information and actionable.
- Big Data: is data so "big" that it is impossible for humans to understand it naturally, or in other words, it is data in a volume and format that can only be analyzed by computer systems.
- Internet of Things (IoT): describes physical objects (or groups of such objects) that are embedded with sensors, processing ability, software, and other technologies that connect and exchange data with other devices and systems over the Internet or other communications networks.
- Web bots or Web crawlers: A Web crawler, sometimes called a web bot, spider, or spider bot, is software created to systematically browse and extract data from the World Wide Web.
- SQL: is the abbreviation of Structured Query Language and is a domain-specific language, used in programming and designed for managing data held in a relational database management system, or for stream processing in a relational data stream management system.
- NoSQL: originally known as "non-SQL" or "non-relational" is a mechanism for storage and retrieval of data that is modeled without the tabular relations used in relational databases.
- Data Wrangling: is the process of transforming and mapping data from one "raw" data form into another format with the intent of making it more valuable for a variety of purposes such as analytics.

- Data Mining: is a process used to search, extract, and analyze data that also involves methods of machine learning, formal linguistics analyses such as textual statistics, and database systems.
- Machine Learning: is the study of computer algorithms that can improve results automatically through experience and that can reproduce human decisions using mathematics and statistics.
- Design Thinking: is a term used to represent a set of cognitive, strategic, and practical processes by which design concepts are developed.
- Design Sprint: is a time-constrained, five-phase process based on design thinking with the aim of reducing the risk when bringing a new product, service, or a feature to the market.

Take Home Message/Key Points

- You can't take good decisions without standardized, timely data.
- Very important aspects of digital agriculture are data collection, transmission, and storage technologies.
- Once data reaches its destination, further decisions will take place (in real time or for future operations improvements).
- Data collection, transmission, and storage are enabling a more agile and efficient farm operation management.

Acknowledgments We would like to thank the funding agencies CAPES, CNPq, and FAPEMIG for their financial support and Maria Dalila Vieira for editing some figures in this chapter.

References

Alves L, Antunes É, Ferreira R, Nacif JA (2019) A mesh sensor network based on bluetooth: comparing topologies to crop monitoring. In: Proceedings of the IX Simpósio Brasileiro de Engenharia de Sistemas Computacionais. SBC, pp 125–130

Barrett SFA II (2020) Systems – synthesis lectures on digital circuits and systems. Morgan & Claypool Publishers, United States

Beroleti P (2019) Projetos com ESP32 e LoRa. Editora Instituto Newton C. Braga, Brazil

Caldas R, Corrêa F, Nacif JA, Roque T, Ruirz L, Fernandes A, da Mata J, Coelho C (2005) Low power/high performance self-adapting sensor node architecture. In: 2005 IEEE conference on emerging technologies and factory automation, p 4, pp 976

GSMA (2021) GSMA association. Available in https://www.gsma.com/. Retrieved in Nov 2021

Gull C, Minkov M, Pereira E, Nacif JAA (2016) Low-cost chlorophyll fluorescence sensor system. In: 2016 VI Brazilian Symposium on Computing Systems Engineering (SBESC). IEEE, pp 186–191

Guzman N, Wied M, Kozine I, Lundteigen MA (2020) Conceptualizing the key features of cyber-physical systems in a multi-layered representation for safety and security analysis. Syst Eng 23(2):189–210

IEC, International Electrotechnical Commission (2008) International standard – quantities and units – part 13: information science and technology, IEC 80000-13

Khattab A, Habib S, Ismail H, Zayan S, Fahmy Y, Khairy M (2019) An IoT-based cognitive monitoring system for early plant disease forecast. Comput Electron Agric 166:105028

Kim W, Lee W, Kim Y (2020) A review of the applications of the Internet of Things (IoT) for agricultural automation. J Biosyst Eng:1–16 45:385–400. https://link.springer.com/article/10.1007/
s42853-020-00078-3

Liao M, Chen S, Chou C, Chen H, Yeh S, Chang Y, Jiang J (2017) On precisely relating the growth
of Phalaenopsis leaves to greenhouse environmental factors by using an IoT-based monitoring
system. Comput Electron Agric 136:125–139

Manyika J, Chui M, Bisson P, Woetzel J, Dobbs R, Bughin J, Aharon D (2015) Unlocking the
potential of the Internet of Things. McKinsey Global Institute. https://aegex.com/images/
uploads/white_papers/Unlocking_the_potential_of_the_Internet_of_Things___McKinsey__
Company.pdf

Mashal I, Alsaryrah O, Chung T, Yang C, Kuo W, Agrawal D (2015) Ad Hoc Netw 28:68–90

Matheus L, Pires L, Vieira A, Vieira L, Vieira M, Nacif JA (2019) The internet of light: impact of
colors in LED-to-LED visible light communication systems. Internet Technol Lett 2(1):e78

Minerva R, Biru A, Rotondi D (2015) Towards a definition of the Internet of Things (IoT). IEEE
Internet Initiat 1(1):1–86

Passe F, Vasconcelos V, Canesche M, Ferreira R (2017) Perspectivas para o uso do Node-Red no
Ensino de IoT. Int J Comput Arch Educ (IJCAE) 6(1):46–51

Pepper R (2019) Cisco visual networking index: global mobile data traffic forecast update,
2017–2022. Cisco technical report, vol 2017, p 2022

Santos B, Silva L, Celes C, Borges-Neto J, Peres B, Vieira MA, Vieira LF, Goussevskaia O,
Loureiro AA (2016) Internet das coisas: da teoria à prática. Short courses SBRC-Simpósio
Brasileiro de Redes de Computadores e Sistemas Distribuídos, vol 31

Tang Y, Dananjayan S, Hou C, Guo Q, Luo S, He Y (2021) A survey on the 5G network and its
impact on agriculture: challenges and opportunities. Comput Electron Agric 180:105895

Vieira LF, Vieira MA, Nacif JA, Vieira AB (2018) Autonomous wireless lake monitoring. Comput
Sci Eng 20(1):66–75

Zhang J, Tao D (2020) Empowering things with intelligence: a survey of the progress, challenges,
and opportunities in artificial intelligence of things. IEEE Internet Things J 8(10):7789–7817.
https://ieeexplore.ieee.org/document/9264235

Zuquim AL, Vieira LF, Vieira MA, Vieira AB, Carvalho H, Nacif JA, Coelho C, da Silva D,
Fernandes AO, Loureiro AA (2003) Efficient power management in real-time embedded systems. In: EFTA 2003. 2003 IEEE conference on emerging technologies and factory automation. Proceedings (Cat. No. 03TH8696). IEEE, pp 496–505

Chapter 13
Data Transmission, Cloud Computing, and Big Data

Hannes Fischer, Jose Vitor Salvi, and Luis Hilário Tobler Garcia

13.1 Introduction

Producing high-quality food in a sustainable and intensive way to quench the hunger of a population that still grows is still a great challenge. To overcome this problem, a higher level of knowledge and control over the various variables that involve decision-making in this complex field is necessary.

In agricultural activities, there are dozens of controllable variables, such as the amount of fertilizer, seeds, and pesticides and the type of equipment to be used, among other things. However, there are also dozens more uncontrollable variables, such as climate, soil biology, and commodity prices. Therefore, extracting valuable information and building strategic knowledge for good decision-making in such a complex reality has become necessary for farmers. An example of a result obtained from this type of strategy is the increase in the average corn yield in recent years in the USA (Fig. 13.1).

Very much technology and human talent have been involved throughout history to reach an average of over 170 bushels of corn produced per acre (11,4 ton/ha) in the USA in 2019. Record-setting farmers can produce over 380 bushels per acre (25,6 ton/ha with the best of genetics, nutrition, and automation and maybe, a bit of luck). If we look at some years ago (Fig. 13.2), we can observe that the greatest advances in agriculture happened in the industrial era, known in agribusiness as modern agriculture and which had as a high point the development of chemical fertilizers, agricultural pesticides, agricultural mechanization, irrigation, highly productive varieties, and the start of information technologies.

H. Fischer (✉) · J. V. Salvi · L. H. T. Garcia
São Paulo State Technology College (FATEC Pompeia Shunji Nishimura), São Paulo, Brazil
e-mail: hannes@fatecpompeia.edu.br; jose.salvi@fatec.sp.gov.br; luishilario@fatecpompeia.edu.br

D. Marçal de Queiroz et al. (eds.), *Digital Agriculture*,
https://doi.org/10.1007/978-3-031-14533-9_13

221

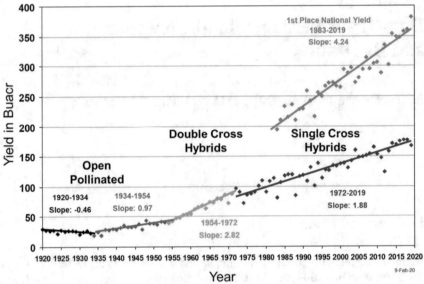

Fig. 13.1 Evolution of corn yield in the USA in the last 100 years. (Source: USDA, NASS, 2020)

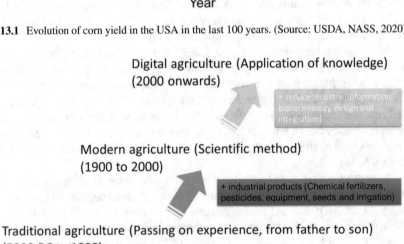

Fig. 13.2 Evolution of agriculture. (Based on personal communication with Dr. Samson Tsou, Professor Emeritus at Ping Tung National University, Taiwan, China)

The next step for a new increase in productivity will depend on a new generation, which will be responsible for deeply understanding modern agriculture and the tools and technologies of digital agriculture. Therefore, the young readers of this work have a mission: prepare to work more and more integrated with professionals in the most diverse areas, because the world is now transdisciplinary and multidisciplinary. And the new things, the innovation, and the future will come from the convergence of many technologies and fields. The professionals of agriculture will be involved with the professionals of computing, IoT, robotics, genomics, and biotechnology, among many other specialties that were not previously imagined. More and more it will be necessary to tear down the walls between people and companies and make efforts to develop important new skills in this new world. Food will be produced in so many ways than we know today, from plant factories to bioreactors and small and large open field farms.

It will be extremely important that this new flexible and highly adaptive professional develops soft skills such as proactivity, autonomy, collaboration, delivery of results, innovation, entrepreneurship, people management, relationship management, and emotions management, among others (also known as M- or T-shaped person). But no less important will be the values that such professionals hold dear, like honesty, humility, and hard work, among others, often not learned in the cradle. And this is connected to a purpose of life that makes sense, a vision that is beyond my own self. Such efforts will take the history of agriculture to a level never seen before, and it is quite possible that the average 400 bushels per acre will be the new standard of productivity in 2040.

The adoption of advanced and unconventional technologies will depend heavily on the engagement of the farmer with professionals that understand the field. It is not uncommon that some skepticism regarding the potential of the technology and its promises to obtain better results in the field has been often observed. And there are many reasons for this distrust: lack of end-user experience, unscrupulous companies that promise more than they can deliver, as well as difficulties in implementing high-tech products and services (reality is way more complex than it seems to be).

Thus, the correct understanding of the technologies is fundamental to avoid financial losses and headaches. It is no accident that large cooperatives and agricultural companies have created teams dedicated to evaluating these innovations and only after proving the value that these technologies add and understanding the best way to use them, do they pass them on to their cooperative members. The use of technologies such as satellite images (Huang et al. 2018), airplanes, drones (Boursianis et al. 2020), most diverse types of sensors (Queiroz et al. 2020), and traps, among other technologies described in the chapters of this book, demands increasing efforts to develop the infrastructure for data transmission in the field and lowering the cost, thus, allowing telemetry to happen in real time. This way, the captured data will be transferred into huge data frameworks (*data lakes*), using *edge computing* and *cloud computing*. Thus, it will be possible to extract value and create information and knowledge that enhance results as never before (you can find more information on this topic in other chapters of this book).

13.2 Data Transmission

13.2.1 Telemetry System

Telemetry is the remote collection of data obtained by sensors or some other source and its automatic transmission to some equipment that receives and makes use of that data. In telemetry we have a module that is responsible for preparing and transmitting the data in the form of a message, and the receiving device can receive this message and extract the data in it, making it available for its proper use. Most agricultural machines are currently equipped with a Controller Area Network (CAN) bus, through which all the data collected by the different systems of the machine are made available on that network. The transmitting device of the telemetry system, once connected to this network, can extract the data of interest generated by the different systems of the machine and then transmit them. Figure 13.3 illustrates a CAN bus of an agricultural tractor.

In the case of a self-propelled sprayer, the bus will allow communication with sensors that are in the machine (engine, transmission, hydraulic system, etc.) and with spray-related sensors (valves, pressure, flow, temperature sensors, and relative humidity). All sensors are connected to the controller modules that can perform specific functions on the equipment, such as controlling the powertrain (diesel engine, transmission, hydraulics) or the vital functions of a sprayer, sending and monitoring parameters to ensure the correct volume of application in a given condition. However, there are modules that do not perform actions but collect and transmit information, through the Internet, to a local server or the cloud. This is the module that will perform the telemetry, that is, that will do the remote collection and transmission of the data (Fig. 13.4).

Fig. 13.3 CAN bus from an agricultural tractor. (Source: John Deere 2018)

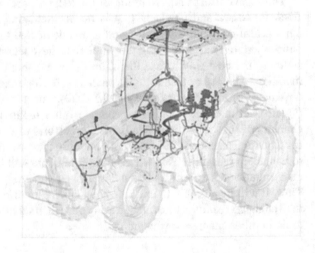

Fig. 13.4 Example of a telemetry module of a self-propelled sprayer. (Source: Adapted from Jacto 2019)

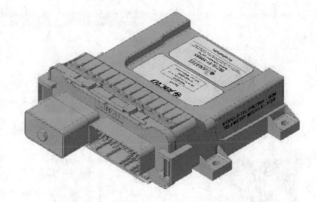

Table 13.1 Collected and transmitted data from a self-propelled sprayer

Operator data	Environmental conditions
Identification	Relative humidity of the air
Running service order	Ambient temperature
Spray status	**Operation status**
Instant fuel consumption	Displacement
Instantaneous speed	In operation
Machine alarms	Stop
Application status	Occurrence
Instantaneous and programmed dose	No precision
Volume and area applied	No connection

Source: Adapted from Jacto (2019)

13.2.2 Data Collection and Transmission

As an example, in self-propelled sprayers, the main information collected is related to operator data, sprayer status, application status, environmental conditions, and operation status. Table 13.1 shows the details of the data components.

Figure 13.5 shows data transmission types of a self-propelled sprayer equipped with a telemetry system. Some manufacturers/model these systems come as default, but sometimes a subscription is necessary to use the full potential. The investment in these cloud subscriptions or virtual platforms is less than 0.4% of the value of the machine.

Figure 13.5 shows an example where you can transmit data in three distinct ways: mobile internet, *Wi-Fi*, or via a *Wi-Fi* collector (outdated technology). In mobile internet transmission (mobile data), data is transmitted whenever a 2/3/4/5G connection is available. The collection and transmission rate varies for each equipment and manufacturer, and in the case of a self-propelled sprayer, the current data collection rate is around 5 s (or when there is any event change – if any nozzle changes state from open to closed, if the quick relief session commands are triggered by the operator, etc.). The transmission is initiated when a certain size in bytes of the collected information is reached, and the data is then sent (when there is

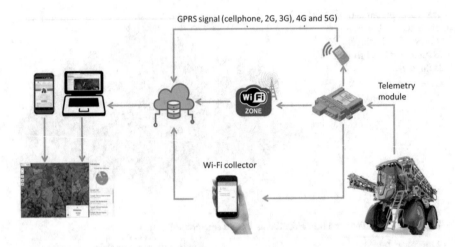

GPRS signal (cellphone, 2G, 3G), 4G and 5G

Telemetry module

Wi-Fi collector

Fig. 13.5 Example of a telemetry system of a self-propelled sprayer. (Source: Adapted from Jacto 2019)

connectivity). When there is no connectivity, the data is stored locally and sent when it is restored. All collected data is sent to the cloud, which can be accessed by the user, through a *tablet, notebook,* or *smartphone.* Finally, it is possible to use the telemetry system in several ways: for the almost real-time monitoring of the machine fleet, for the management of agricultural operations,[1] and/or for the analysis of operational maps and application of inputs.

13.2.3   Farm Connectivity

The collection and transmission of data allow the farm management team to monitor the equipment so that they can know in real time the application conditions and *status* of the equipment. In about 10 min (depending on the manufacturer), the data is already processed, and the operational and application maps are available. If the farm manager needs to monitor his fleet in real time, the company will need connectivity throughout the area. Connectivity in the field is increasingly accessible and with better quality. Several initiatives supported by agribusiness companies have projects to stimulate and enable Internet access.[2]

However, when the self-propelled sprayer is operating in a property or location where there is no mobile data coverage, the data will continue to be collected every 5 s, as mentioned in the previous section, and stored in the telemetry module.

---

[1] Delays of up to 5 min even with optimal connectivity are common, and this will tend to decrease in the future with new communication technologies such as 5G, for instance.

[2] One of these projects in Brazil is ConectarAGRO – https://conectaragro.com.br

Because the data is encoded and compressed, memory consumption is reduced. In this case, the stored data can be transmitted in two ways: the first is a situation in which there is wireless network coverage (*Wi-Fi*) near the place where the equipment is stored, then the module will be able to connect to the configured network and transmit the data to the cloud. The transmission of the data is maintained even if the operator shuts down the equipment, since the module remains in operation for a certain time, to complete the transmission. The second way is when there is no wireless coverage (*Wi-Fi*); in this case, the collection and transmission of the data may occur through a *smartphone* of the operator or person in charge. In this situation, the telemetry module generates a *Wi-Fi network* so that the *smartphone* can connect. When you establish the connection, the data will be transferred to your mobile phone and finally be transmitted to the cloud when an Internet connection is available. This way, there is also no loss or corruption of data, making it a practical, safe, and effective way to map agricultural operations.

13.3 Cloud Computing

As we have seen in the previous section of telemetry, in the past all data was stored on a local computer of the farm, but in many cases, data was not collected at all. Data is now becoming increasingly important, and there is a growing movement to store them in the cloud. Cloud computing is a method of outsourcing on-demand computing resources through the concept of distributed computing, in which thousands of computers connected to each other via the Internet behave virtually as a single source of resources to be consumed by users. An illustration of this is the way how electricity supply is currently outsourced, in which users do not have to worry about how it is produced or distributed. Consumers can simply use the resources and services offered in the "cloud" and pay for what they have consumed.

Cloud computing allows the user to simply use it for data storage (saving data collected from machines, photos, documents, etc.), use its processing power (to run apps, calculations or run machine learning algorithms, etc.), and have development environments to run applications without having to worry about how they work. Cloud computing would not have been possible without the Internet. The cloud is an abstraction that hides a complex infrastructure connected to the Internet and is a computing concept in which resources are provided as services, allowing users to access them without knowledge or control over the technologies behind these servers. The cost of computing power and Internet bandwidth and speed has been decaying exponentially in the past decades, and we have recently witnessed the convergence of all these technologies, making it always faster and cheaper.

One of the main features of cloud computing is the ability to grow, or scale, as the users need more processing power; it will easily adjust its available resources and expand its computational power as well as its storage capacity. Today, with the standardization of everything that involves cloud computing, the environments have become much more user friendly, making it easy to use cloud computing services from distinct vendors. However, we still have some work ahead on the standardization of the agricultural value chain, and this is a crucial step toward the digitalization and explosion of new services that the next generation of farmers will experience.

It is possible to mention, among the major suppliers of this type of technology, companies like Alibaba, Amazon Web Services, Microsoft Azure, Google Cloud, IBM Cloud, and Oracle Cloud, but there are also many small businesses disputing this same marketspace.

Cloud computing has allowed startups to bloom in a variety of areas, making it possible to access an infrastructure that was expensive and inaccessible to companies in their early stages. Today, someone with a good idea can quickly prove value and, often using free tools, develop a mobile app and create solutions previously unimaginable. The cloud computing suppliers many times offer some free services till a certain cap is reached (data transferred or processed, etc.). This enables startups to test the feasibility of their solution, since deployment costs are dramatically reduced.

13.3.1 Example of Cloud Computing in Agriculture

The agricultural sector is undergoing a futuristic revolution,[3] for example, the difficulty in controlling weed infestation is leading to the development of robots. The robot detects and eliminates weeds in the crop, as well uses several embedded sensors (accelerometers, electronic compasses, gyroscopes, cameras, GNSS,[4] LiDAR, etc.) to indicate the precise location of the pests. This is made possible thanks to a computer with high processing power and machine-learning algorithms for weed identification. Finally, the system can apply the correct herbicide to the weeds.

In most practical cases, the robot's connectivity would not be good enough to perform cloud processing, at least for now. That way, the robot has a sufficient and powerful computer onboard, with a cost in the range of $200. Currently, the automotive industry is technologically at the forefront and makes use of several super-powerful computers, which are in the same price[5] range, to use them in their driver assistance systems.

---

[3] https://www.ecorobotix.com/en/ Yes https://www.swarmfarm.com

[4] Mobile phones like the Xiaomi Mi 8 have static positioning errors of the order of 30 cm and dynamic errors in the 2 m range (with Broadcom BCM47755 chipset using two GNSS frequencies, the GALILEO E1/E5a and GPS L1/L5). The use of E5 and L5 frequencies was impossible before 2015, as there were no satellites that used the bands before – https://spectrum.ieee.org/tech-talk/semiconductors/design/superaccurate-gps-chips-coming-to-smartphones-in-2018

[5] Version 2.5 used by Tesla in 2016 was able to process 4 TFLOPS (4 trillion operations per second), and this corresponded to processing 200 images per second; today this value is close to 400 TFLOPS (https://en.wikipedia.org/wiki/Tesla_Autopilot), processing over 2.000 higher definition images per second.

With increased connectivity in the field, the weed eliminator robot will be able to send all data to the cloud. The processing capacity will be greater, and more complex models of machine learning can be applied. Another advantage will be the possibility of this model being constantly updated, generating even better results over time. Herbicide suppliers will have real-time information about the level of infestation in that region and will be able to program themselves to serve their customers.

The trend in agriculture is the combination of embedded or edge computing with cloud computing. Think of insect detection traps, in which a super simple computer like a raspberry Pi[6] can count the number of insects present in the trap and send this information to the cloud. There, this data can then be processed further and generate information for the farmer, such as alerts.

13.4   Big Data

In this new era of digital agriculture (Fig. 13.2), also known as the "Age of Knowledge," the volume of data generated daily by various segments of society[7] is so large that traditional visualization tools have become inefficient in identifying relevant information for correct management and good decision-making. Data has been available in various agricultural operations and has been stored by farmers. All this data from all these several different operations is known as "Small Data," which is important to understand each production process, but when associated with other data sources can become way more important.

To make the agricultural production process more efficient and increase profitability in the field, several initiatives have taken place to incorporate sensors and actuators into agricultural machinery and their implements. These devices made operations more accurate and capture data throughout all the use of the machinery, showing details about the operation and the equipment used.

These data, associated with other data sources that were previously underutilized, allowed for the creation of information that, when properly used, may assist the decision-making of the various professionals involved in agricultural production. But today, the amount of data generated after a day of operation of self-propelled spraying equipment equipped with high technology goes beyond the limits of traditional data analysis tools and hinders good governance by the managers.

It was found that data from completely different sources, when associated, can show correlations and relationships of cause and effect previously unperceived and

---

[6] Model 4 with 2 GB of RAM costs $35 in 2020, and a computer with the same performance will cost less than $1 in 2030 – https://www.raspberrypi.org

[7] Every 10 years, the amount of data generated and stored has increased more or less by a factor 100.

of great value for good decision-making, but associating data from various sources is more complicated than it seems.

To better understand this, imagine putting together in a single place data related to a rural property, where we would have the history of the soil's biological, chemical, and physical characteristics; market data for several commodities; price prediction in the future market; fuel prices; prices of inputs (pesticides, fertilizers, energy, water, etc.); weather data (air speed, air direction, air temperature, relative humidity, solar incidence, rainfall, forecasts); pest history; planted varieties; among others. In a situation like this, we have more variables than the human brain can process at once, and even using traditional data analysis tools, the decision matrix would have so many dimensions that we would be unable to extract all the potential information that this volume of data could offer.

The output is the adoption of new technologies, tools, and methods of computing, present in an area known today as Big Data. This term describes a new area of science that uses theories, techniques, tools, and methods from various areas of knowledge, including Computer Science, Information Science, Mathematics, and Statistics. This new area of science became part of human history from the popularization of the Internet and only managed to grow because of the evolution of the processing power of computers.

The life cycle of a big data study closely resembles the scientific research method adopted by most researchers worldwide. A study in Big Data also proposes research on a hypothesis or event that occurs on a farm and aims to draw answers and conclusions from this study. However, in this case, the raw material is the data, and it is worth highlighting the importance of the quality of the data generated and stored, since only good data will generate useful information, upon which knowledge can be built and decisions taken. This will make the agribusiness better positioned in the world scenario.

It is necessary for the farmer to be aware of the importance to collect correct and reliable data, free of noise that compromises its quality, store it properly, and exploit it with the correct techniques and tools.

From a larger picture, it is possible to mention several tools that are part of this world, among them are the most modern languages of software development, the development of "Internet of Things" (IoT) devices, used in capturing data through sensors, the development of software known as web bots or web crawlers, used for data capture on the Internet. In this same context are the most modern forms of data storage, among them relational databases (SQL), non-relational databases (NoSQL), including all distributed and cloud storage technologies, as well as block storage that has links to each other (Blockchain). With the selected data, we have the application of data preprocessing techniques, including data cleaning and normalization (Data Wrangling), mathematical processing techniques and data transformation, which is associated with statistical analysis (Data Mining), Data Visualization, development of predictive models (Machine Learning), and other tools. Figure 13.6 illustrates the steps in the life cycle of a Big Data study.

To better understand this ongoing process that involves Big Data research, see below a brief explanation of each step.

Fig. 13.6 Lifecycle of a
Big Data study. (Source:
Authors of this chapter,
2020)

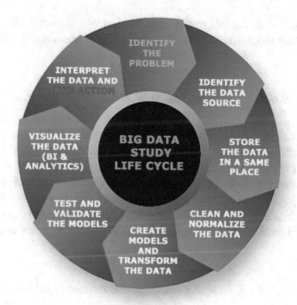

13.4.1 The Identification of the Problem

At this stage, it is necessary to discuss the main purpose of the study. This will happen only if there is a good integration of the various experts in each technology area involved with the business segment to be studied. Among the skills needed for the success of this stage are soft skills like the ability to work as a team, creativity, and awareness of the problem to be solved. It is worth mentioning that methodologies such as "Design Thinking" or "Design Sprint" can be used to assist in the process of defining the hypothesis to be studied. If there is no problem to be solved, one should not spend resources on this activity. Always have an answer to the question "What problem do you want to solve through this study?"

13.4.2 Identifying the Source of the Data

The terms that summarize this step are "Volume," "Variety," and Veracity. It is necessary to collect as much data from as many different sources as possible, while always maintaining the objective of capturing data that can contribute to the preparation of the desired responses and that has the best possible level of quality. It is important to mention that in situations where the user does not have high-quality data, it is possible to improve it using some processing techniques, interpolation, and even artificial intelligence. This way, it is possible to fill the gaps in the data sample and make them better prepared for the next stages of the study. But this will not work for all types of data sources. So, the better the quality the easier the next steps become. Around 20% of the time is spent at this stage.

13.4.3 Concentration of Data in One Place

At this stage, the volume and variety characteristics of the data are considered to choose the best database management strategy (SQL or NoSQL), whether on-premises or cloud storage, etc. Among the forms of data storage, we also have Blockchain technology, which in the future might have great importance for creating ways to ensure reliable traceability of agribusiness products. In this technology, data is stored on a huge set of computers connected through the Internet, where an algorithm encrypts the data and prevents stored data from being changed or erased.

13.4.4 Cleaning and Standardizing Data (Data Wrangling)

In this step, several techniques are used to ensure that data from different sources is stored in the same format. Here irrelevant data for the study can be removed and techniques that fill important gaps in the data may be applied. Today, due to the lack of standardization, a data scientist loses 60% of the time in this step, a very annoying but crucial step. This time will be reduced in the future with better standards in place, so a great improvement in the workflow will be possible.

13.4.5 Creation of Models and Transformation of Data

In this step, the data is ready for one of the most important phases of the study, when it is transformed using algorithms and mathematical models, as well as machine learning techniques, and thus high added value information is generated, which with specific field knowledge will lead to good decision-making. Around 10% of the time is spent at this stage.

13.4.6 Testing and Validation of Models

In this step all work is tested in some of the data and, if the models generated in the previous steps are explaining the data well, it might be applied to the rest of the database. These results are then checked to confirm whether the answer to the problem that is desired to be solved was achieved. If this did not happen, a return to the previous step will be necessary, with the goal to adjust the parameters of the models and algorithms until satisfactory results are achieved (sometimes this might not even happen, then everything must restart from the beginning). Around 5% of the time is spent at this stage.

13.4.7 Data Visualization, Analysis of Results, and Decision-Making

At this last moment of the study, the data and mathematical models have already been validated and the information generated can be used for the creation of control panels, or dashboards, containing graphs, tables, and statistical data, from which analysis and decision-making will take place.

13.5 Concluding Remarks

Visionary entrepreneurs were always first to adapt the most advanced technologies, even if the price point was not there yet. And now, for more than 20 years, farmers are using this type of service, monitoring several parameters in their farms (from soil, plant, to climate). They, for example, determine the optimal irrigation time and amount (see Chap. 10 on Digital Irrigation) with these technologies. Sensors developed by agricultural companies are getting cheaper and cheaper, thus resulting in a revolution in generated data. This data must then be transmitted to the cloud so that it can be processed and combined with many other types of data, which can be obtained from satellite images, radar, commodity prices, and social media trends, among others, delivering knowledge (which is the transformation of information with the acquired experience of people) so that the farmer can make the best decision.

However, challenges will arise, such as hackers trying to hack and steal what is valuable to the farmer, which is his data. There will also be trust issues. How will farmers trust all these companies that will have access to their valuable data? But nothing will stop the next generation of farmers, they will continue to reinvent themselves faster and faster, because they will be better prepared and much more connected with those who have the knowledge – and will be less hierarchical (Antle 2019). This new generation will know when to enter a new business venture and when to get out of it. They will not lose their time with lost battles. A revolution in agriculture is believed to occur in the next 10 years, with the adoption of unprecedented digitization. In general, everything that is digitized ends up getting much cheaper, faster, and better. See how the automotive industry is changing rapidly with digitization, for example, cars receive performance upgrades over the cloud and soon will no longer need drivers. Passengers of these vehicles will become mostly consumers of digital services and will pay for them. And how will this impact the agribusiness world? What will it be like in the future? (Dönitz et al. 2020). Certainly, the agribusiness value chain will be reinvented and be completely different from what we know today (Coble et al. 2018).

Reading this book might eventually frighten the readers a bit, but they will not be alone in the pursuit of the use of this noble item called "data," because this task should be undertaken with the help of professionals who are being prepared to give them the necessary support. Multidisciplinary professionals with several additional skills are being trained, including professionals trained in *Big Data* or Data Science focused on agricultural problems, who can assist you in the challenges that are ahead.

Complex and powerful solutions will arise and most sold as a service to the farmer. But remember, to take advantage of this data revolution, you must have internal competencies that have the right knowledge or partner with someone that has it, otherwise, you might be fooled or waste your time.

There is much potential for overall improvement in the agribusiness value chain, and the technologies described in this book will help the readers to explore them. The important thing is to participate in the flow of knowledge and not stay on the sidelines of information. But remember, ruptures in the market will not come necessarily from all these wonder technologies, but from the customer. He will tell you how and what to produce.

Welcome to the new normal, in which agriculture is no longer just machines or agronomy, or… just computing! It is all of this and increasingly integrated and well-orchestrated.

Abbreviations/Definitions

- AI: Artificial intelligence (AI) refers to the simulation of human intelligence in machines that are programmed to think like humans and mimic their actions. The term may also be applied to any machine that exhibits traits associated with a human mind such as learning and problem-solving.
- Big data: very large sets of data more complex data sets, especially from new data sources. These data sets are so voluminous that traditional data processing software just can't manage them. But these massive volumes of data can be used to address business problems you wouldn't have been able to tackle before.
- Blockchain: a system in which a record of transactions made in bitcoin or another cryptocurrency is maintained across several computers that are linked in a peer-to-peer network.
- Cloud computing: is the delivery of different services through the Internet, including data storage, servers, databases, networking, and software.
- Edge computing: is the practice of capturing, storing, processing, and analyzing data near the client, where the data is generated, instead of in a centralized data-processing warehouse.
- IoT: the networking capability that allows information to be sent to and received from objects and devices (such as fixtures and kitchen appliances) using the Internet. The Internet of Things really comes together with the connection of sensors and machines.

- LiDAR: stands for Light Detection and Ranging, is a remote sensing method that uses light in the form of a pulsed laser to measure ranges (variable distances) to the Earth.
- Machine learning: a type of artificial intelligence in which computers use huge amounts of data to learn how to do tasks rather than being programmed to do them. Machine learning makes it possible for computing systems to become smarter as they encounter additional data.
- SQL: Stands for structured query language – full structured query language, computer language designed for eliciting information from databases.
- Telemetry: the use of radio waves, telephone lines, etc., to transmit the readings of measuring instruments to a device on which the readings can be indicated or recorded.

Take Home Message/Key Points

- You cannot take good decisions without standardized and timely data.
- Big data is a very important aspect of digital agriculture.
- The volume of collected data from equipment and sensors is huge, requiring special computing capabilities to be processed.
- Big data enable farmers to be more agile and efficient in real-time decision-making.

References

Antle JM (2019) Data, economics and computational agricultural science. Am J Agric Econ 101(2):365–382. https://doi.org/10.1093/ajae/aay103

Boursianis AD, Papadopoulou MS, Diamantoulakis P, Liopa-Tsakalidi A, Barouchas P, Salahas G, Karagiannidis G, Wan S, Goudos SK (2020) Internet of things (IoT) and agricultural unmanned aerial vehicles (UAVs) in smart farming: a comprehensive review. Internet of Things:100187. https://doi.org/10.1016/j.iot.2020.100187

Coble KH, Mishra AK, Ferrell S, Griffin T (2018) Big Data in agriculture: a challenge for the future. Appl Econ Perspect Policy 40(1):79–96. https://doi.org/10.1093/aepp/ppx056

Dönitz E, Voglhuber-Slavinsky A, Moller B (2020) Agribusiness in 2035 – farmers of the future. Fraunhofer ISI

Huang Y, Chen Z, Yu T, Huang X, Gu X (2018) Agricultural remote sensing big data: management and applications. J Integr Agric 17(9):1915–1931. https://doi.org/10.1016/S2095-3119(17)61859-8

Jacto MA (2019) Manual Técnico Uniport 4530. Pompeia, 587 p

John D (2018) Manual do operador: monitor GreenStar 2630. Moline, 412 p

Queiroz DM, De Freitas Coelho AL, Valente DSM, Schueller JK (2020) Sensors applied to digital agriculture: a review. Rev Ciênc Agron 51

Chapter 14
Machine Learning

Domingos Sárvio M. Valente, Daniel Marçal de Queiroz, and Gustavo Willam Pereira

14.1 Introduction

The term "Machine Learning" can be understood as the ability of computers to learn from data. Basically, it can be divided into classification and regression algorithms. Classification generates categorical (or discrete) results, while regression generates predictions of continuous values. Machine Learning algorithms can also be classified into supervised and unsupervised learning. In supervised learning, for each input sample, the expected result in the output, also known as the label, is also presented to the algorithm.

There are several supervised learning algorithms, such as K-Nearest Neighbors (K-NN), Linear, Polynomial and Logistic Regression, Support Vector Machine (SVM), Decision Tree, Random Forest (RF), LightGBM, XGBoost, and Artificial Neural Networks (ANN). In unsupervised learning, only input samples are presented to the algorithm, and the data used to fit (train) the models are not labeled. Thus, the algorithm is expected to classify the data into groups (clusters). Among the algorithms, it is possible to mention the clustering algorithms (cluster such as K-Means, Fuzzy K-Means, and Hierarchical Clustering (agglomerative, divisive)). There are other classifications for Machine Learning algorithms such as semi-supervised learning, reinforcement learning, online learning, batch learning, model-based learning, and instance-based learning.

D. S. M. Valente (✉) · D. Marçal de Queiroz
Universidade Federal de Viçosa, Viçosa, Brasil
e-mail: valente@ufv.br; queiroz@ufv.br

G. W. Pereira
Instituto Federal de Educação Ciência e Tecnologia Sudeste de Minas Gerais,
Campus Muriaé, Brasil
e-mail: gustavo.willam@ifsudestemg.edu.br

Machine Learning models can use categorical and numerical data (binary, integer, real) from different types of sensors. These data are used to train (fit) the model. The dataset is composed of observations (samples or instances) and features. Observations are organized into rows in the dataset, and features are organized into columns. This format is the standard used in Machine Learning algorithms.

One of the benefits of digital agriculture is the obtaining of large volumes and varieties of data from the agricultural system. These data should be processed to extract useful information in decision-making. Machine Learning models use these data to make predictions, such as the level of severity of a disease, needed dose of fertilizer, water depth to be applied, rainfall forecast, and pest infestation forecast (Ferentinos 2018; Liakos et al. 2018; Kamilaris and Prenafeta-Boldú 2018; Jha et al. 2019).

This chapter will address common programming languages and Integrated Development Environments (IDEs) used in Machine Learning. After that, the main steps of data preprocessing will be described. Next, a classification problem will be proposed and solved from the beginning, using the Random Forest algorithm. In the end, the basic principle of operation of the Decision Tree and Random Forest algorithms will be presented.

14.2 Programming Languages

The way to pass instructions to computers is through programming languages. There is a vast list of programming languages that can be used, such as *C*, *C++*, *Java*, *Visual Basic*, *PHP*, *R*, *Python*, and *Fortran*, among many others. In general, the programming logic does not change from one language to another. Understand the logic, then it will be like riding a bike, something you never forget, and that is what really matters. However, if the intention is to work with data science and Machine Learning, we strongly recommend *Python* or *R* language, because they have a vast set of modules (ready-to-use algorithms, also known as libraries) that assist programmers in developing a project.

In this chapter we will present some examples using the *Python* programming language. *Python* is an easy-to-understand, simple language with clear and straightforward syntax. A lot of things can be done with a few command lines, combining performance and practicality. Furthermore, there is a large community of programmers who are developing modules (complex algorithms ready to be used) that can be imported and executed in your code.

Python offers scientific modules, for vector and array operation, image processing, data science, optimization, statistics, graph plotting, graphical interface, and Machine Learning (*Scikit-Learn*), which is the goal of this chapter. By using the modules, *Python* becomes a language that will mediate the access to the functions of the selected module. This reduces the amount of code and increases productivity. Another advantage of *Python* is flexibility, which means that one can use it to create programs for multiple platforms (*Windows, Linux, macOS, Android,* and *iOS*).

Python is an object-oriented language, and languages of this type make it much easier to develop computer programs and integrate them with modules made available by the community. In addition, you can make small sets of instructions (scripts) for specific software programs such as *QGIS* and *ArcGIS*. These scripts automate repetitive tasks that would take a very long time if performed manually.

14.3 Basic Tools

The interpreter for *Python* can be installed through a download from the official website (https://www.python.org), where there is also extensive documentation on the language. However, this path can be rather time-consuming, since only the language interpreter and its default libraries are installed. A list of the standard library modules can be found at https://www.python.org/doc/. Therefore, you should probably install a distribution called *Anaconda*.

Anaconda has the function of preparing the programming environment. This means that, when installing *Anaconda*, in addition to the *Python* interpreter (chosen version), the IDE's *Spyder* and *Jupyter Notebook* as well as the *Anaconda Prompt* will also be installed. *Spyder* and *Jupyter* are IDEs where *Python* codes are written. With the use of *Anaconda Prompt*, it will be easy to install additional modules in *Python*. In addition to these software programs, when installing *Anaconda*, basic *Python* modules will also be installed, such as *NumPy* (module for vectors and arrays); *SciPy* (module for mathematics, science, and engineering); and *MatPlotLib* (module for plotting), in addition to several others. To install *Anaconda*, just go to the official website (https://www.anaconda.com) and download a distribution with *Python 3*. Once *Anaconda* is installed, *Spyder* or *Jupyter Notebook* will be available to write *Python* codes. Writing code in these software programs makes it much easier to visualize, fix (debugging), and run scripts. As an exercise, look on the internet for information on how to install *Anaconda* and how to use *Spyder, Jupyter Notebook,* and *Anaconda Prompt.*

The next step is to learn the *Python language.* It is necessary to understand and know the common and special types of variables (list, tuples, and dictionaries) of *Python*, control structure, selection and repetition, functions, and an introduction of classes and objects. With respect to classes and objects (object-oriented programming), try to understand the concepts and use of classes. Probably it will not be necessary to implement classes in your programs, but you will want to make use of them.

14.4 Data Preprocessing

The first thing to do before you start running Machine Learning algorithms is data preprocessing. Preprocessing is the preparation of data to be used in Machine Learning algorithms. Although many do not give due attention to preprocessing, it can be said that it is the most important step in the design of a model and probably the most time-consuming step. After all, if you enter "poor data" into your model, the output will definitely be "poor data." In addition, it is in the preprocessing of data that knowledge (insights) is extracted to create a better Machine Learning model.

Among the tasks that must be performed in preprocessing, it is possible to mention the following: (a) visualization of the data, (b) verification and filling of missing data, (c) verification and removal of discrepant data (outliers), (d) deletion of redundant data, (e) standardization or normalization of data, (f) treatment of categorical data, and (g) reduction of dimensionality of training data. To perform these tasks, it is recommended to use the modules *NumPy*, *Pandas*, *MatPlotLib*, and the specific module of Machine Learning *SciKit-Learn*.

14.4.1 Main Modules

The *NumPy* module performs operations with arrays. For having hundreds of mathematical functions for operation with vectors and arrays, *NumPy* is the basis for numerical computing in *Python*. *NumPy* was developed in *C*, so it is very efficient in calculations. During the development of Machine Learning models, some *NumPy* functions are used.

The *Pandas* module is one of the most important tools for data scientists. Powerful Machine Learning tools can attract all the attention, but *Pandas* can be considered the backbone of most Machine Learning models. *Pandas* is used to clean, transform, analyze, and visualize data from different angles. This will help you decide the path to be followed in data modeling. For now, understand *Pandas* as if it were a *Microsoft Excel* that will run within *Python*. With *Pandas* you can open datasets with different extensions, such as CSV (data file in text format, where columns are separated by commas). These files will open in *DataFrame* variables. A *DataFrame* is similar to a spreadsheet with rows and columns. As an exercise, open a CSV file (text file with fields separated by comma or semicolon) using *Pandas*. A line of code using the *read_csv()* function will be required. Tip: Copy the CSV file to the same folder where the *Python* code file is saved. As an exercise, do a search on other functions of *Pandas*. When you understand *Pandas* well, move on.

MatPlotLib is a module for creating plots. Plots can be highly customized. With it you can create line plots, scatter plots, bar plots, and histograms, among several others. *MatPlotLib* is considered a lower-level module, especially when the intention is to create more complex plots. However, you can find many examples of code for constructing plots. In addition to *MatPlotLib*, there are other modules for creating plots, such as *Seaborn*. *Seaborn* is a module built based on *MatPlotLib*, so it is a module that can build sophisticated plots with few lines of code. Create some plots on your *Spyder*. Advance after you know how to generate at least one line plot (*plot* function), scatter plot (*scatter* function), bar plot (*bar* function), and histogram (*hist* function).

Scikit-Learn is the Machine Learning module for *Python*. *Scikit-Learn* contains preprocessing tools, such as processing of missing data (NAN, from *not-a-number*), processing of categorical data, and standardization and normalization of data, as well as many others. There are various Machine Learning algorithms for regression, classification, clustering, model selection, and reduction of dimensionality.

The following preprocessing methods will be described below: data standardization, missing data processing, and categorical data processing. But remember, many other preprocessing tasks can be performed; it will depend on your problem.

14.4.2 Data Standardization

Many Machine Learning algorithms use some distance criteria to generate the models. If the features have different numerical scales, each feature will impact to a greater or lesser degree the modeling as a function of that scale. This means that the importance of some feature will not be defined by its actual importance, but rather by the numerical scale, and this is not correct. See the following example: the *altitude* feature has a numerical variation between 800 and 1200, and the *NDVI* feature (normalized difference vegetation index) varies between 0.2 and 0.7. Now imagine that the response variable has low sensitivity to *altitude* variation and, on the other hand, great sensitivity to *NDVI* variation. It is expected that a good model will be generated and that *NDVI* will be more important. However, depending on the Machine Learning algorithm used, if the *altitude* and *NDVI* values are not standardized for the same numerical scale, for example, with mean 0 and standard deviation 1, the accuracy of the model will be low. This will occur because the *altitude* with the numerical scale from 800 to 1200 will "swallow" the entire variation of *NDVI*, with the scale between 0.2 and 0.7. Therefore, for some algorithms, it is recommended that data be standardized or normalized.

In some cases, it is recommended to perform standardization even in binary variables (0 and 1). Also, even if the algorithm does not require standardization, it is often important to "help" the algorithm find optimal values of the functions. Below is an example of the *Scikit-Learn* preprocessing module being used for data standardization. The code below was extracted from the official website of *Scikit-Learn* (Pedregosa et al. 2011). We recommend that you read this documentation.

```
4 from sklearn import preprocessing
5 import numpy as np
6 X_train = np.array([[1., -1., 2.],
7                     [2., 0., 0.],
8                     [0., 1., -1.]])
9 X_scaled = preprocessing.scale(X_train)
10
11 print(X_scaled)
```

In line 4, the *preprocessing* module is imported from the *Scikit-Learn* module. In line 5, the *NumPy* module is imported and renamed to *np* (this feature is important to avoid typing long identifiers). In line 6, a 3 × 3 array is created; the result is stored in the *X_train* variable. In line 9, the *scale* function of the *preprocessing* module of *Scikit-Learn* is called. The result of the operation in line 9 is stored in a variable named *X_scaled*. In line 11, the *X_scaled* array is requested to be presented (printed). Figure 14.1 shows the code implemented in *Spyder*. To run the code, simply press the F5 hotkey or click the green arrow icon, located in the top toolbar of *Spyder* (Fig. 14.1). You can run only part of the code. To do this, select the part of the code that you are interested in running, then click F9 or click the third icon directly from the green arrow (black arrow). Running specific pieces of code is important for finding and fixing erros (debugging).

After the code is run, the variables *X_train* and *X_scaled* will be created on the computer's memory. They can be viewed in the variable explorer of *Spyder* (Fig. 14.1). If you double-click any of the variables, the values stored in the variable will be displayed, as shown in Fig. 14.2.

In Fig. 14.2, the *scale* function is calculated for each column (feature) individually. This means that the algorithm calculated the mean and standard deviation of column 0, and then applied the transformation to the entire column. Similarly, the same transformations were performed for the other columns (features).

In terms of Machine Learning, the best way to perform preprocessing would be using the *StandardScaler* class of the *Scikit-Learn* module. Standardization is

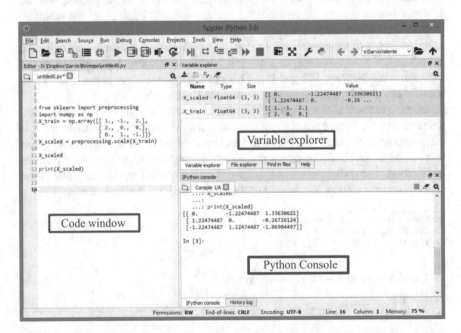

Fig. 14.1 *Spyder* interface with code window, variable window, and *Python* console window. (Source: Authors)

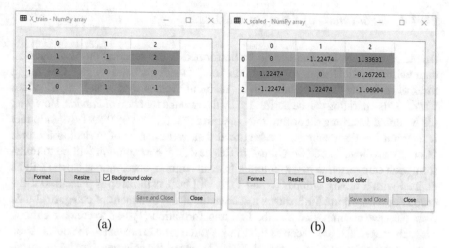

Fig. 14.2 (**a**) Result stored in the *X_train* variable, (**b**) result of the transformation of the *X_train* variable stored in the *X_scaled* variable. (Source: Authors)

performed based on the mean and standard deviation of the data of each column (feature). The standardization equation is as follows:

$$X_{new} = \frac{x - mean(x)}{standard\ deviation(x)}$$

In addition to standardization, there are several other types of data transformation. *StandardScaler* is the most common and most widely used. For example, a normalization can be performed using the *MinMaxScaler* class. In this case, the features will be transformed to values between 0 and 1, for example.

14.4.3 Treatment of Missing Data

A quick fix to solve the missing data problem (NAN) could be to remove the entire row that had the missing data. To do this, simply use the *Pandas dropna* function. This function removes rows or columns (depending on the parameters passed in the function) that contain missing data. Removing the entire row from the dataset that has some missing feature may lead to the loss of important information. Imagine that in your dataset you have 100 observations (100 rows) and 10 features (columns). If each feature has one missing data in different rows, as much as 10% of the dataset will be lost. Thus, to avoid the loss of information, one approach is to estimate the missing data. *Pandas* offers functions for filling gaps (*fillna*). You can find some simple examples in the *Pandas* documentation.

14.4.4 Treatment of Categorical Data

Another preprocessing that should be performed is the substitution of categorical data with numerical data, for example, "low," "medium," and "high." One might think of replacing "low" values with 0, "medium" values with 1, and "high" values with 2. This operation can be performed by the *replace* function of *Pandas*. However, the Machine Learning algorithm may interpret that 0 is worse than 1, when in fact it may not be. For example, if categorical data were names of agricultural crops: "Corn," "Soybean," and "Sorghum," in this case, the best solution will be to transform each category into a new column. In each column, observations (rows) will be defined with a binary number (0 and 1). It will be 1 if the crop is present, 0 if the crop is absent. Figure 14.3 shows a schematic of the treatment of categorical data from the *crop* column. See that the first transformation (1) was to create a column for each crop. Each row receives the value 1 if the crop is present in the observation (row), and 0 if the crop is not present. Then (2), one of the columns must be removed. It will always be necessary to remove a column because it will remove redundant data. See in the example in Fig. 14.3 that only two columns are sufficient to obtain three different classes because we have the combinations (1, 0), (0, 1), and (0, 0). To perform this preprocessing in our dataset, use the *get_dummies* function of *Pandas*. To remove a column, you can use the *drop* function. *Scikit-Learn* offers classes for the treatment of categorical variables; depending on your problem, it may be an interesting solution.

14.5 Machine Learning from the Beginning

A classification problem will be solved in this topic using a dataset provided by Dias (2020). The data were obtained in an experiment to evaluate tomato varieties resistant to late blight (a disease caused by an oomycete that attacks tomato plants). The experiment comprised 132 plots, and each plot had a tomato variety that could or could not be resistant to the disease. Thus, in this exercise, the objective will be to

index	Crop
1	Corn
2	Soybean
3	Corn
4	Sorghum
5	Sorghum
6	Corn

$\xrightarrow{1}$

index	Corn	Soybean	Sorghum
1	1	0	0
2	0	1	0
3	1	0	0
4	0	0	1
5	0	0	1
6	1	0	0

$\xrightarrow{2}$

index	Corn	Soybean
1	1	0
2	0	1
3	1	0
4	0	0
5	0	0
6	1	0

Fig. 14.3 Categorical data preprocessing scheme, in (1) each category is transformed into a new column, in (2) a column is removed

develop a Machine Learning model to classify the plots that showed high or low severity of disease attack for the last day of evaluation in the experiment.

The model will be trained using data of severity assessment in each plot observed by the specialists. Evaluation scores above 40% of severity were called class 1, otherwise class 0. Thus, a dataset was created with one *Class* column and 132 rows (observations). Each row in the *Class* column received values 0 and 1.

The dataset to be used in the Machine Learning model was obtained from images taken by a multispectral camera coupled to a UAV (Unmanned Aerial Vehicle). Images of the experimental area were obtained on four different dates. For each UAV image, the plots were clipped and converted into new images (images of the plot). For each image of the plot, the blue (B), green (G), red (R), *RedEdge* (RE), and infrared (IR) bands were extracted. Based on the spectral bands, the *NDVI* vegetation index was calculated. Then, filtering was performed. The pixel of the clipped image that had *NDVI* value greater than 0.60 was used to calculate other vegetation indices; otherwise, the pixel was discarded. See the scheme of the entire flowchart for data preprocessing in Fig. 14.4.

For this example of application, from the bands filtered in each image of the plot on the four dates of area imaging, 20 average vegetation indices were generated for each plot (*NDVI_d28, SAVI_d28, GNDVI_d28, MCARI1_d28, SR_d28, NDVI_d0, SAVI_d0, GNDVI_d01, MCARI1_d01, SR_d01, NDVI_d04, SAVI_d04, GNDVI_d04, MCARI1_d04, SR_d04, NDVI_d08, SAVI_d08, GNDVI_d08, MCARI1_d08, SR_d08*). Thus, the dataset consists of 132 rows (plots of the experiment) and 21 columns (5 vegetation indices during 4 days of observation and the class). This dataset was saved as a comma-separated text file (CSV type).

Source for downloading the data:

https://github.com/sarviovalente/dataset_tomato/blob/main/dataset_tomato.csv

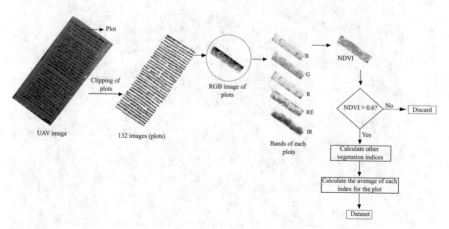

Fig. 14.4 Preprocessing flowchart to generate the dataset for training the Machine Learning model. (Source: Authors)

14.5.1 Prepare the Dataset

First, you must open the dataset in *Python*. This dataset has 132 observations, 20 features, and the *Class*. In the code below, the dataset is imported in *Spyder*.

```
09 # import the necessary libraries
10
11 import pandas as pd          # import pandas module
12 import numpy as np           # import numpy module
13 import matplotlib.pyplot as plt # import matplotlib module
14
15 # open csv file of dataset
16
17 data = pd.read_csv('dataset_tomate.csv')
```

In line 17 of the code, the *Dataset* is imported using the *read_csv* function of *Pandas*. When you click the *data* variable, in the variable window of *Spyder*, a window with the contents of the variable will open, as shown in Fig. 14.5. The *data* variable is of the *dataframe* type, which is a special type of variable of *Python*.

In the *dataframe data*, there are several features. The *dataframe* contains 132 observations (organized in rows) and 21 columns. The features were named with the vegetation index followed by the day of data collection. In the *Class* column (*target*), the class of each observation is defined. As the class of each observation is known, then a supervised classification model will be generated.

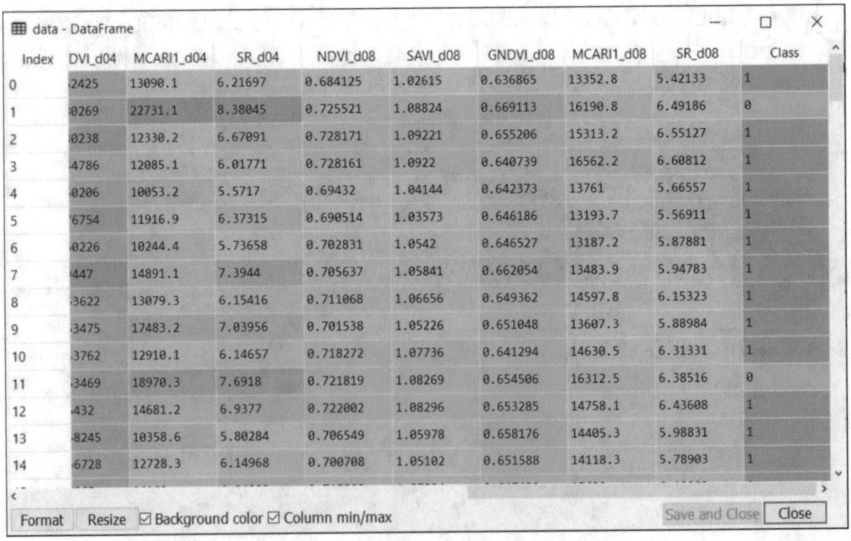

Index	DVI_d04	MCARI1_d04	SR_d04	NDVI_d08	SAVI_d08	GNDVI_d08	MCARI1_d08	SR_d08	Class
0	2425	13090.1	6.21697	0.684125	1.02615	0.636865	13352.8	5.42133	1
1	0269	22731.1	8.38045	0.725521	1.08824	0.669113	16190.8	6.49186	0
2	0238	12330.2	6.67091	0.728171	1.09221	0.655206	15313.2	6.55127	1
3	4786	12085.1	6.01771	0.728161	1.0922	0.640739	16562.2	6.60812	1
4	0206	10053.2	5.5717	0.69432	1.04144	0.642373	13761	5.66557	1
5	6754	11916.9	6.37315	0.690514	1.03573	0.646186	13193.7	5.56911	1
6	0226	10244.4	5.73658	0.702831	1.0542	0.646527	13187.2	5.87881	1
7	447	14891.1	7.3944	0.705637	1.05841	0.662054	13483.9	5.94783	1
8	3622	13079.3	6.15416	0.711068	1.06656	0.649362	14597.8	6.15323	1
9	3475	17483.2	7.03956	0.701538	1.05226	0.651048	13607.3	5.88984	1
10	3762	12910.1	6.14657	0.718272	1.07736	0.641294	14630.5	6.31331	1
11	3469	18970.3	7.6918	0.721819	1.08269	0.654506	16312.5	6.38516	0
12	432	14681.2	6.9377	0.722002	1.08296	0.653285	14758.1	6.43608	1
13	8245	10358.6	5.80284	0.706549	1.05978	0.658176	14405.3	5.98831	1
14	6728	12728.3	6.14968	0.700708	1.05102	0.651588	14118.3	5.78903	1

Format   Resize   ☑ Background color ☑ Column min/max   Save and Close   Close

Fig. 14.5 Dataset stored on a *dataframe*. (Source: Authors)

First, you should extract only the necessary information from the dataset. As the dataset has 20 features, you can use all of them in the Machine Learning model. Using the *data* variable, the X variable (feature array) is created, and with the *Class* column of the *data* variable, the *y* variable (*label*) is created. See the code below to create the X array and the y vector.

```
20 #selection of variables that will be used in the model
21
22 cols = ['NDVI_d28', 'SAVI_d28', 'GNDVI_d28', 'MCARI1_d28', 'SR_d28',
23         'NDVI_d01', 'SAVI_d01', 'GNDVI_d01', 'MCARI1_d01', 'SR_d01',
24         'NDVI_d04', 'SAVI_d04', 'GNDVI_d04', 'MCARI1_d04', 'SR_d04',
25         'NDVI_d08', 'SAVI_d08', 'GNDVI_d08', 'MCARI1_d08', 'SR_d08']
26
27 X = data [cols]
28
29 y = data[['Class']]
```

The only preprocessing to be performed will be the preparation of the dataset to enter the algorithm and the separation of data for training and testing. The *Random Forest* algorithm will then be used to train the classification model. Finally, cross-validation will be used for the evaluation and selection of models. In this dataset, there are no missing data and no categorical data. It will not be necessary to standardize the data because the *Random Forest* algorithm will be used.

14.5.2 Training and Test Data

Machine Learning models often have a great ability to fit to the data. So, to check how accurate the model is it is necessary to take some precautionary measures. Imagine that you generated a model with 98% overall accuracy (correct classifications against total observations). Will the model show the same accuracy with new observations? What can guarantee that? The best way then is to use new data (not used in training). Since the model is under development, and there is usually no time or condition to collect new data, the procedure is to separate some of the data for the test. Some test data should be saved and should not be used until the modeling is finished. If the test data is used during model creation, and not in the end, a biased model may be created. This will occur because you will tend to choose a model that is best for the test data. So, separate the test set and keep it securely "saved" until the modeling is finished. In the end, you will test your model with new data, which are the data you have reserved for testing. To separate the X and y dataset into training data (which will be used in modeling) and test data (data for final testing), you can use the following code:

```
31 #division of the dataset into training and test
32
33 #import the function to divide the dataset
34 from sklearn.model_selection import train_test_split
35
36 #divide the dataset into training and test
37 X_train, X_test, y_train, y_test = train_test_split(X, y, test_size =
0.20, random_state = 0)
```

In line 34, the *train_test_split* function that is used to randomly split the dataset was imported. In line 37, the function is performed. In this function, you enter the dataset X and y; then, the percentage of data that you want in the test dataset is passed, in this example 20% (*test_size* = 0.20). The last parameter is *random_ state* = *0* to always generate the same dataset of training (*X_train, y_train*) and test (*X_test, y_test*) every time you run this program. If you remove *random_state* = *0* from the function, every time you run the code, another dataset of training and test is generated. After running the above code, the variables *X_train, y_train, X_test*, and *y_test* will be available in the variable window of *Spyder*. Click on them and see the results.

14.5.3   Machine Learning Model

Now that the training (*X_train* and *y_train*) and test (*X_test* and *y_test*) variables are defined, the Machine Learning model will be generated. In this example, the *Random Forest* algorithm will be used. As the idea of this topic is to present the steps of modeling in Machine Learning, the principle of operation of the *Random Forest* algorithm will not be presented in detail, which will be done in future topics. To generate the *Random Forest* model, the following lines of code will be required:

```
40 # import the Random Forest class for classification
41 from sklearn.ensemble import RandomForestClassifier
42
43 # create the classifier object
44 clf = RandomForestClassifier()
```

The above code is slightly different from those previously used. In this case, a class was imported and not a function. By convention, classes have capital letters in their name. So, in line 41 the *RandomForestClassifier* class was imported (it is not a function as in line 37 of the previous code). This class is for supervised classification using the *Random Forest* algorithm. Unlike functions, classes cannot be used directly, only in specific situations, when the class has static attributes or methods. In general, you must create an object using the class as a reference. As an analogy, the design of a car is a class, the object is the car created with all the characteristics of engine, gearbox, steering wheel, doors, color, etc. Back to the code, in line 44 the object called *clf* (it could have the name you want) was created. The object was created using the constructor of the *RandomForestClassifier()* class. In object-oriented languages, such as *Python*, every method that has the same class name is called a constructor and its primary purpose is to instantiate an object, that is, to allocate memory to the object. Run the commands, and if no error occurs (you can view it in the *Spyder* console window), the *clf* object will be created.

The object will now be used to fit the classifier based on the training data. This will be done with a command line.

```
46 # fit the model
47 clf.fit(X_train, np.ravel(y_train))
```

In line 47, the *clf* object is fitted with the *fit* function. The parameters of the function are the training data (*X_train*, *y_train*). Run the code. Done! The *clf* classifier object is trained and can be used to make predictions. In the code below, the *clf* classifier is used to predict the training data (*X_train*). Thus, it is possible to compare and verify if the model obtained good accuracy within the training data. Test data remains "stored."

```
49 # predict y_train with X_train data
50 y_pred_train = clf.predict(X_train)
```

In line 50, the prediction was performed using the *predict* function of the *clf* object. To make the prediction, only the training data (*X_train*) were passed as parameters. The prediction result will be stored in the *y_pred_train* variable. Figure 14.6 shows the first 10 rows of *y_train* and *y_pred_train*. Compare the predicted and actual results. In the first 10 rows the accuracy was 100%. But did the same occur for all the data? One way to verify accuracy for all the data is to use an evaluation function, for example, overall accuracy. There are several ways to evaluate a classifier. For each problem, there is a more appropriate method of evaluation; seek information about false positive, false negative, confusion matrix, precision, recall, f1 score, and kappa coefficient, among others.

In the code below, the *score* function is used to determine the overall accuracy considering all observations of the training data (*y_train*) in relation to the predicted data (*y_pred_train*).

```
52 # calculate the overall accuracy for the training data
53 score_train = clf.score(X_train, y_train)
```

In the *score* function, the following parameters were passed: the *X_train* array, which will be used to make predictions in the *score* function, and the *y_train* vector, which represents the observed values. The function will take care of making the prediction and then performing the comparison between the predicted and observed data. The accuracy result will be stored in the *score_train* variable. The result can be viewed in the variable window of *Spyder*. The result was 1.0 (100.0% of hits). Is

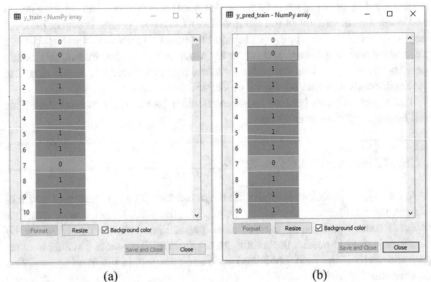

(a)   (b)

Fig. 14.6 Comparison between observed values and values predicted by *Random Forest*: (**a**) Observed classification of the class for the dataset and (**b**) Classification predicted by *Random Forest* classifier of the class for the dataset. (Source: Authors)

that really what it is? If new data are collected or simply if test data are used in the model, will the hit rate be the same? Has there not been an overfit of the model? After all, the predicted data were used in the training. For this and other reasons, the next topic is extremely important in Machine Learning: cross-validation.

14.5.4 Cross-Validation

Machine Learning algorithms can have great fitting capacity, depending on the parameters defined in them. In addition, some Machine Learning problems have a very high relationship between the number of features and observed data. This may lead to the generation of overfitted models. Some authors recommend at least 25 observations for each feature in the dataset. Obviously, this number of observations will be easier or more difficult to obtain depending on the problem. In the tomato problem, presented in this chapter, 20 features were used, so in theory, we should have 500 observations. One solution would be to reduce the number of features to a maximum of five. Therefore, in Machine Learning problems, it is interesting to reduce the dimensionality or select the most important features to build the model. In addition, the selection of the most important variables is one of the most relevant steps in the construction of models. Simpler and more robust models can be built. To proceed with the selection of characteristics, cross-validation is used.

In addition to the selection of features, Machine Learning algorithms have several parameters that can be defined at the time of the creation of the classifier or regressor object. In Machine Learning, we call these parameters hyperparameters because they must be selected by the user. In the *Random Forest* algorithm, the main hyperparameters are *n_estimator* and *max_depth*. The meaning of each of these hyperparameters will be explained in future topics. Understand these hyperparameters as if they were the degree of a polynomial. The hyperparameter *n_estimator* could assume values such as 10, 20, 100, 500, etc., and similarly the hyperparameter *max_depth* could assume values such as 3, 4, 5, 10, 100, etc. Depending on the combination of hyperparameters, it is possible to obtain a better model. Cross-validation is used to choose the best one. As an exercise, look in the documentation of the *RandomForestClassifier* algorithm of *Scikit-Learn* for the hyperparameters that can be defined.

Cross-validation consists of using the training data (*X_train* and *y_train*) and dividing them, for instance, into five sets (these sets are called *k-folds*). Then, in this example, the algorithm should use sets 1, 2, 3, and 4 to perform the fitting, and set 5 will be used to calculate accuracy. Then, the process is repeated, now using sets 1, 2, 3, and 5 for fitting and set 4 to calculate accuracy. This routine will be repeated five times.

In the problem in question, cross-validation will be performed with five divisions (*k-folds* = 5). At the end of the cross-validation, five values of accuracy will be obtained, and their average should be calculated for the model. Then, the values of the hyperparameters of the *Random Forest* model are changed, cross-validation is performed again, and a new average of accuracy will be obtained. The model with the best average of accuracy will be the selected model. This same procedure can be used to select the features.

In the code below, the classifier object (line 56) was created with the hyperparameters *n_estimators* = 100 and *max_depth* = 10. The *random_state* = 0 was also inserted to generate the same model every time the *script* is run again.

```
55 # create the classifier object
56 clf = RandomForestClassifier(n_estimators = 100, max_depth = 10, random_state = 0)
```

Then, the code to perform the cross-validation is inserted.

```
58 # import the function for cross-validation
59 from sklearn.model_selection import cross_val_score
60
61 # perform cross-validation of the object with training data
62 scores = cross_val_score(clf, X_train, y_train, cv = 5)
```

In line 59, the cross-validation function (*cross_val_score*) was imported. In line 62, cross-validation will be performed. In the *cross_val_score* function (line 62) the classifier object (*clf*), the training data (*X_train* and *y_train*), and the number of *k-folds* (cv = 5) were passed as parameters. The results will be stored in the *scores* variable. The *scores* variable will be a vector with five accuracies. The result can be seen in the variable window of *Spyder*. A code is inserted in the sequence to calculate the average of the accuracies using *NumPy*.

```
64 # use NumPy to calculate the mean of the scores vector
65 mean = np.mean(scores)
66 print('cross-validation score:', mean)
```

In line 65, the calculation of the mean is performed, and the result is stored in the *mean* variable. In line 66, the *mean* variable is printed. The result of the mean should be 91.42% (that is, if you defined the *random_state = 0* in line 56). As an exercise, change the values of the hyperparameters of line 56 and run again (lines 55–66). Use this mechanism and try to find the combination that will generate the highest mean. This method of searching for optimal hyperparameters seems a little annoying. In fact, *Scikit-Learn* provides a class to automate the search (*GridSearchCV*). We suggest you look for more information. A more efficient method of searching for optimal hyperparameters is with the *Scikit-Optimize* module (https://scikit-optimize.github.io/stable/). We strongly suggest using this module, as it will not test all possibilities according to *GridSearchCV*, but will explore optimal points. In the *Scikit-Optimize* module, there is the *BayesSearchCV* class, which is compatible with *GridSearchCV*.

Once the model is selected, you should evaluate how it behaves with the test data. See the code below.

```
68 # fit the selected model with training data
69 clf.fit(X_train, y_train)
70
71 # predict y_test with X_test data
72 y_pred_test = clf.predict(X_test)
73
74 # calculate overall accuracy for test data
75 score_test = clf.score(X_test, y_test)
```

In line 69, the classifier is fitted using all training data (*X_train* and *y_train*), considering that the classifier object was defined in line 56. In line 72, the prediction is performed with the test data (*X_test*). In line 75, the overall accuracy is calculated with the test data (*X_test* and *y_test*). The result of overall accuracy for the test data should be 96.29%.

In an attempt to improve the performance of the classifier one can include new features or create features from the combination of existing ones. The creativity and knowledge of the scientist may be decisive in defining the new features. Another way to improve the accuracy of the model would be to use other Machine Learning algorithms or a combination of algorithms (*Ensemble Learning*).

As an exercise, try using the *Support Vector Machine* algorithm to classify the severity in the problem presented. In the code below the algorithm is imported, and the classifier object is created with two hyperparameters, *C* and *gamma*.

```
77 # import the SVC class for classification
78 from sklearn.svm import SVC
79
80 # create the object defining two hyperparameters (C and gamma)
81 clf = SVC(C = 100, gamma = 10)
```

To use the *SVC* (*Support Vector Machine for Classification*), you will need to standardize the training data (*X_train*). Likewise, it will be necessary to standardize the test data (*X_test*) to make the prediction. This is very important since test data are not available, and they do not exist in theory. You must create a standardization object and fit it (*fit*) with the training data. Then, perform the standardization (*transform*) using only training data (*X_train*). See the code below.

```
83 # standardize training data
84 from sklearn.preprocessing import StandardScaler
85 scaleX = StandardScaler()              # create a standardization object
86 scaleX.fit(X_train)                    # fit to training data
87 X_train_new = scaleX.transform(X_train) # standardize training data
```

In line 84, the *StandardScaler* class is imported. In line 85, the *scaleX* object is created using the *StandardScaler* class. In line 86, the *scaleX* object is fitted with the training data (*X_train*). Finally, in line 87 the training data (*X_train*) are transformed using the *transform* function of the *scaleX* object. The result is stored in the new variable *X_train_new*. This variable is the one that will be used to fit the *SVC* model and perform the cross-validation.

To make the prediction with the test data, the test data (*X_test*) must be standardized. As the *scaleX* object has already been created and fitted, simply apply the transformation to the test data using the same *scaleX* object. See the code below.

```
89 # standardize test data
90 X_test_new = scaleX.transform(X_test)
```

14.6   Machine Learning Algorithms

Machine Learning algorithms typically have a great ability to fit the training data. Therefore, if you do not tune some hyperparameters, your model may become over-fitted (*overfitting*). Thus, it is important to have basic knowledge of algorithms so that you understand how hyperparameters affect their fitting. This topic describes the working principle of the *Decision Tree* and *Random Forest* algorithms.

14.6.1 Decision Tree

The *Decision Tree* algorithm constructs regression or classification models in the form of a decision tree structure. It divides a dataset into smaller subsets. The new subset of data is expected to have more homogeneous results than the original set, that is, it is expected to have less impurity. Basically, the algorithm selects a feature and determines a threshold for the feature that best distinguishes the dataset. The feature that generates the subset with the least impurity will be placed at the top of the tree (root node). Thus, the subdivision of the nodes of the tree will occur until it arrives at a final decision (leaf node), in which there will be no more subdivisions. The subdivision may be interrupted when the generated result has greater impurity than the original data. Another stop criterion in the subdivision may be the definition of some hyperparameter, for example, the maximum depth of the tree (*max_depth = 5*).

The end result of the *Decision Tree* algorithm is a tree with decision nodes and leaves. On the leaf, the decision is considered the basis for the statistics of its content. If it is a classification problem, the decision could be made based on the highest probability of each class. If it is regression, the result may be a simple average of the leaf data.

An example of a decision node is presented in Fig. 14.7. In this example, the *Iris Flower* dataset and the petal length feature were used. The optimal threshold for petal length is defined based on an impurity score. There are some ways to calculate impurity, such as Gini coefficient, entropy, ID3, or CART. Figure 14.7 shows the Gini coefficient calculated for two situations, with a petal length threshold equal to 2.5 (Fig. 14.7a) and a petal length threshold equal to 5.1 (Fig. 14.7b).

The Gini coefficient is calculated based on the following equation:

$$\text{Gini} = 1 - \sum_{c=1}^{n} p_c^2$$

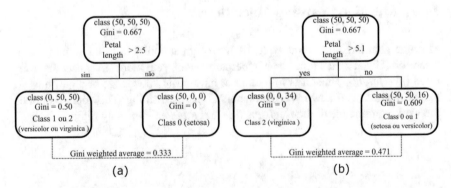

(a)   (b)

Fig. 14.7 Subdivision of the node for the petal length feature for two thresholds (**a**) 2.5 cm and (**b**) 5.1 cm. (Source: Authors)

Where c represents a given class and n is the number of classes. In the example used, we have three different classes, and p is the probability of each class occurring on the node. Here is an example of the calculation of the Gini coefficient for the root node in Fig. 14.7.

$$\text{Gini} = 1 - \left(\frac{50}{150}\right)^2 - \left(\frac{50}{150}\right)^2 - \left(\frac{50}{150}\right)^2 = 0.667$$

Only two different thresholds for petal length were presented in Fig. 14.7: 2.5 cm (Fig. 14.7a) and 5.1 cm (Fig. 14.7b). The algorithm should determine the optimal threshold for each feature. The threshold of 2.5 cm generated subdivisions with Gini of 0.50 and 0. The Gini coefficient of the division is calculated based on the weighted average. For Fig. 14.7 the calculation is determined as follows:

$$\text{Gini} = 0.5 \frac{100}{150} + 0 \frac{50}{150} = 0.333$$

The same calculation was performed for the threshold of 5.1 cm, and it generated a Gini coefficient of 0.471. In this case, the algorithm will decide that the best threshold will be 2.5 cm. The calculation will be performed for all features and, in the end, the feature that shows the lowest Gini coefficient (lowest impurity) will be defined as the root node of the tree. In the sequence, new subdivisions can be performed, or leaves can be defined.

As already mentioned, during modeling it is always interesting to adjust the hyperparameters to limit the model and thus avoid overfitting. In the decision tree there are several hyperparameters. An important hyperparameter is the depth of the tree (*max_depth*). Cross-validation can be used to define the best values for hyperparameters.

An advantage of the *Decision Tree* algorithm is the ease of interpretation of the generated tree, that is, it is easy to extract knowledge from the dataset. One disadvantage is that the *Decision Tree* algorithm is greedy, which means that it creates a tree by placing on top the feature that "best" defines the dataset. However, this is not always the best solution. As an exercise, implement the *Decision Tree* algorithm for the problem of the severity of tomato disease. Make modifications to the hyperparameters and see what happened.

14.6.2 Random Forest

Machine Learning algorithms generally perform better when we use more than one algorithm for decision-making. You can also achieve better accuracy by generating multiple models with the same algorithm, modifying the data source. It is based on this strategy that the *Random Forest* algorithm works. Basically, it generates multiple *Decision Tree* models based on the random choice of part of the training data.

The *Random Forest* algorithm has the following steps: (1) construct a subset of random data from the training data; (2) randomly select a subset of features from the subset in step 1; (3) create a decision tree from the subset; and (4) Repeat step 1, 2, and 3 to generate several different trees. The number of trees generated (*n_estimator*) is a hyperparameter that can be defined. The final classification will be defined for the class with the highest number of votes. If it is regression, the result can be obtained by the average of the predictions of each tree.

The *Random Forest* algorithm has a great capacity for learning (fitting). Thus, to avoid overfitting, it is recommended to limit its degrees of freedom. To do this, some hyperparameters of the algorithm must be defined. Two important hyperparameters are the number of trees generated (*n_estimator*) and the maximum depth of the tree (*max_depth*). The choice of the best hyperparameters should be made based on cross-validation. However, as previously seen, the *Random Forest* algorithm uses only part of the data to generate a tree, so part of the training data is left out of the modeling (this set is known as *out-of-bag*). Thus, depending on the amount of data available for training, the *out-of-bag* set can be used to evaluate the model. Search the *Scikit-Learn* documentation for more information.

Random Forest is one of the most important algorithms of *Scikit-Learn*. However, more information about other *Scikit-Learn* algorithms, such as *Logistic Regression*, *Naïve Bayes*, and *K-Nearest Neighbors*, should be sought. Also, although it was not addressed in this chapter, try to create regression models. For regression, you can use the following algorithms: *Linear Regression*, *Support Vector Regression*, *Decision Tree*, and *Random Forest*. Then you can proceed to artificial neural networks. For artificial neural networks we recommend the *Keras* module. *Keras* was developed based on the *TensorFlow* module (developed by *Google*). Artificial neural networks can be used to create powerful models of classification and regression. In addition, convolutional neural networks (*deep learning*) can be used to classify digital images. Many scientific studies have shown good models for identifying pests and diseases in agriculture using *deep learning*. The use of artificial neural networks is a more advanced topic, so we suggest you start with the basic models, which, despite being basic, can solve complex problems.

Abbreviations/Definitions

Module: Python package or library with well-defined objectives, which has a set of classes to perform its functions.

Integrated Development Environment (IDE): composed of a text editor, interpreter, or compiler (depending on the language being used) and code debugger that helps the analyst/programmer in writing and analyzing a computer program being developed.

Scripts: the set of instructions/commands written in a programming language, which when executed perform a certain automated task.

Features: set of attributes or characteristics used by Machine Learning algorithms to perform training. Also known as independent variables (X) of the dataset.

Label: dependent variable, class (y), label. Value to be found from training the
   model using the features as input data.

Supervised learning: the training data provided to the algorithm includes classes (y)
   or labels.

Unsupervised learning: the training data provided to the algorithm does not include
   classes (y) or labels.

Insiders: information that is important, relevant, and produced by the Machine
   Learning model.

Normalized Difference Vegetation Index (NDVI): Vegetation index calculated based
   on red and near-infrared bands.

Unmanned Aerial Vehicle (UAV): drone used with a sensor (camera) to collect images.

Ensemble Learning: the new model is based on the combination of Machine
   Learning models. For example, a random forest model is a combination of the
   decision tree models.

Overfitting: consists of developing a machine learning model with excellent performance on training data, but when subjected to new data, it generates poor results.
   We often say that the model "memorized" and did not "learn."

Confusion Matrix: matrix structure used to evaluate classification models. Each row
   represents an actual class, and each column represents a predicted class. Correctly
   classified values are on the main diagonal – True Negative (TN): class 0–0; True
   Positive (TP): class 1–1. Misclassified values are off the main diagonal – False
   Negative (FN): Class 1–0; False Positive (FP): Class 0–1.

Precision: metric used to determine the accuracy of positive predictions from a
   machine learning model, based on the values of the confusion matrix. Precision
   is calculated using the following equation:

$$\text{precision} = \frac{TP}{TP + FP}$$

Recall: also known as sensitivity or rate of true positives that are correctly detected
   by the machine learning model. The recall is calculated using the following
   equation:

$$\text{recall} = \frac{TP}{TP + FN}$$

f1-score: Metric that combines precision and revocation into a single metric. The
   f1-score is the harmonic mean of precision and recall. The model will only get a
   high f1-score if the revocation and precision are high. The f1-score is calculated
   using the following equation:

$$f1-\text{score} = \cfrac{2}{\cfrac{1}{\text{precision}} + \cfrac{1}{\text{recall}}}$$

Take Home Message/Key Points

- The term "Machine Learning" can be understood as the ability of computers to learn from data.
- To work with data science and Machine Learning, we strongly recommend *Python* or *R* language.
- Machine Learning models often have a great ability to fit the data. So, to check how accurate the model is, it is necessary to use new data (not used in training).
- Spend time generating new features for your model.

References

Dias FO (2020) Modelagem de tendências espaciais na seleção de linhagens de tomateiro resistentes à Phytophthora infestans (Mont.) de Bary. Dissertação de Mestrado, Universidade Federal de Viçosa, 49p

Ferentinos KP (2018) Deep learning models for plant disease detection and diagnosis. Comput Electron Agric 145:311–318

Jha K, Doshi A, Patel P, Shah M (2019) A comprehensive review on automation in agriculture using artificial intelligence. Artif Intell Agric 2:1–12

Kamilaris A, Prenafeta-Boldú FX (2018) Deep learning in agriculture: a survey. Comput Electron Agric 147:70–90

Liakos KG, Busato P, Moshou D, Pearson S, Bochtis D (2018) Machine learning in agriculture: a review. Sensors 19(8):2674

Pedregosa F, Varoquaux G, Gramfort A, Michel V, Thirion B, Grisel O, Blondel M, Prettenhofer P, Weiss R, Dubourg V, Vanderplas J, Passos A, Cournapeau D, Brucher M, Perrot M, Duchesnay E (2011) Scikit-learn: machine learning in python. J Mach Learn Res 12:2825–2830

QGIS Development Team (2017) QGIS geographic information system. Open Source Geospatial Found. Project

Chapter 15
Platforms, Applications, and Software

Andre Luiz de Freitas Coelho, Thiago Furtado de Oliveira, and Maurício Nicocelli Netto

15.1 Introduction

This chapter aims to show some of the platforms, applications, and computer programs (software) that are being employed in digital agriculture. These three computational tools have contributed to the optimization of the food production system in a global scenario. That is, these tools assist the farmer in the management of crops, facilitating data analysis and decision-making aimed at optimizing the production system. The optimization of the production system based on information obtained from the tools used in digital agriculture aims to rationally use production inputs. This can lead to a reduction in production costs and/or an increase in yield, resulting in greater profitability for the farmer. This optimization also makes it possible to reduce the environmental impacts on the agricultural production system, with the rationalization of the use of chemical inputs and reduction of pollutant emissions by internal combustion engines.

An important computational tool for digital agriculture is the Geographic Information System (GIS). In general, GIS is a set of integrated software, hardware, data, and users that make it possible to collect, store, manage, and analyze georeferenced data, enabling the production of information from their use. Currently, some GIS software available for use are ArcGIS, QGIS, gvSIG, SAGA GIS, and GRASS GIS, among others. These computer programs are used in various applications, such as mapping, telecommunications, urban and transportation planning, hydrological analysis, environmental impacts, and agriculture. There are also GIS

A. Luiz de Freitas Coelho (✉) · T. F. de Oliveira
Federal University of Vicosa, Vicosa, Brazil
e-mail: andre.coelho@ufv.br; thiago.oliveira@ufv.br

M. N. Netto
MONAGRI Consultancy, Mato Grosso, Brazil
e-mail: mauricionicocelli@gmail.com

software developed specifically to meet the demands of digital agriculture. These programs have been referred to as Farm Management Information Systems (FMISs).

According to Saiz-Rubio and Rovira-Más (2020), FMIS are tools with the function of assisting the farmer in various tasks, from planning and execution of operations to management of documents. Thus, these systems have the capacity to automate data acquisition and processing, monitoring, planning, decision-making, and management of documents and operations. They may also include functions for records and reports of yield, profits and losses, scheduling of agricultural operations, weather forecasting, soil nutrient tracking and field mapping, farm accounting, inventory management, and documentation of employee contracts.

In this chapter, some FMISs used in digital agriculture will be presented. Some computational tools with specific functions, such as the acquisition and processing of data from sensors onboard unmanned aerial vehicles, and applications for mobile devices are also presented. Finally, the challenges and the future of the use of computational tools in global agriculture are presented.

15.2 Telemetry and ISOBUS

One of the capabilities of FMISs presented is the automation of data acquisition. To understand this automation of data acquisition, we will first address the concepts of telemetry and ISOBUS.

The term "telemetry" has a Greek origin, in which *tele* means "remote" and *metron* means "measure." Telemetry allows data obtained by various sensors to be transmitted to a storage server (called "the cloud") using the Internet. These data are presented to the farmer in the FMIS in ways that facilitate their interpretation or analysis such as maps, graphs, tables, and indexes. More concepts about telemetry were present in the chapter "Internet of Things for Agriculture."

In addition to automating data acquisition and transmission, another benefit of telemetry is that data can be viewed by the farmer remotely and in real time. Thus, telemetry makes it possible to monitor all agricultural operations, such as sowing, spraying, fertilizer application, and harvesting, while they are being performed. Monitoring makes it possible to achieve greater operational capacity and higher efficiency of operations since the farmer/manager has a broad and detailed knowledge of the operation. This knowledge allows the farmer to make decisions such as reallocating machines and changing operational parameters. As a direct consequence of the adoption of telemetry, there is a reduction in the costs of operations.

The advancement and wide dissemination of telemetry of agricultural operations have only been possible with the implementation and adoption of ISOBUS standard in tractors and agricultural machinery. When adopting ISOBUS standard, manufacturers of different agricultural machines use the same communication and control standard, which enables the communication between the different components of the tractor and of the machines (Figs. 15.1 and 15.2).

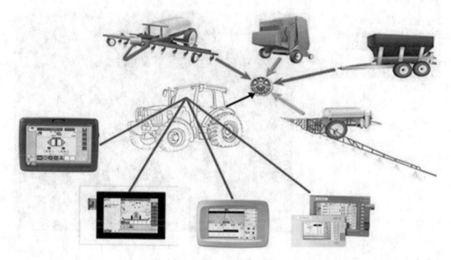

Fig. 15.1 ISOBUS: Standardized communication protocol between different agricultural machines. (Source: AgLeader 2020)

Fig. 15.2 Standard connector used in equipment that adopts ISOBUS. (Source: Kverneland Group 2022)

ISOBUS is defined in ISO 11783 standard, adopting the CAN 2 (Controller Area Network) communication protocol, developed by the German company BOSCH. ISOBUS allows access to the data generated by all sensors of the tractor and the machines attached to it without difficulties and restrictions, such as encryption added by manufacturers. Using equipment from the machine manufacturer itself or from third parties, the data made available by the tractor and machine can be transmitted to the server, hence being accessible to the farmer remotely and in real time. An example is the Climate FieldView platform, which uses a drive connected to the tractor's ISOBUS connector (Fig. 15.3) to collect the data provided by the tractor and machine and send them to the company platform.

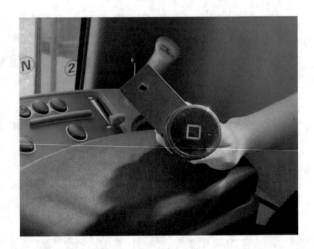

15.3 Farm Management Information Systems

Saiz-Rubio and Rovira-Más (2020) presented 36 platforms of FMISs and their relevant characteristics (Table 15.1). Some of these platforms have global dissemination, such as those developed and made available by big companies that manufacture tractors, agricultural machines, and equipment used in agriculture. Others have dissemination in a country or region of the country and are developed by small- and medium-sized companies.

In addition to the FMIS presented by Saiz-Rubio and Rovira-Más (2020), there are others developed by the bigs of the agricultural sector. The Fuse platform, belonging to AGCO Corporation, a group formed by the brands Challenger, Fendt, GSI, Massey Ferguson, and Valtra, has resources for the management of the entire farm, in all agricultural operations, from planting, through cultivation and harvesting, to the storage of grains. For example, the AgCommand and Connect tools are telemetry software programs that provide on the platform machine data such as location, vehicle status, hours, and performance information.

The platforms made available by companies that manufacture agricultural pesticides include BASF's Xarvio and Singenta's CropWise. The Xarvio platform is composed of the following solutions: Scouting and Field Manager. The Xarvio Scouting solution makes it possible to identify weeds, nitrogen nutrition, and leaf damage as well as recognize diseases from images captured with a cell phone. Xarvio Field Manager, on the other hand, is intended for the management of plots from planting to harvesting, as in satellite image-based biomass maps, history of yield maps, and climate information, besides allowing the generation of digital weed maps.

The CropWise Protector platform contains features that assist the farmer in decision-making, through graphs and visual analysis, digital monitoring of operations, analysis of application efficiency, and monitoring of field teams. CropWise Imagery in turn uses satellite, airplane, and UAV (Unnamed Aerial Vehicle) images

Table 15.1 Some FMISs used in digital agriculture

Platform name, company, and country	Relevant features
ADAPT, AgGatekeeper, USA	Input/output translator to manage data among controllers, field equipment, and farm management information system in an adequate format. Open-source system offered at no cost for developers to adopt into their proprietary systems
AGERmetrix, AGERpoint, USA	Crop data and analytics platform with a mapping interface. Able to scan and collect high-resolution crop data through LiDAR and other collaborative techniques. Permits taking data on mobile devices
AgHub, GiSC, USA	Independent solution by a cooperative. Collect and securely stores data. Data can be shared with trusted advisors
Agrivi, Agrivi, United Kingdom	Weather, field mapping, plan inventory. Crop, machinery, and personnel management (notifications and reports). Web-based and mobile versions
Agroptima, Agroptima, Spain	Mobile App as an electronic notebook to record field activities, products applied, workers implied, working time, or machinery usage. Data can be downloaded on Excel, and safely stored in the cloud
AgroSense, Corizon, Netherlands and Spain	Open source. Work done, field data, and timetables can be shared with contractors or employees. Automate importing and interpreting performed tasks via ISOBUS. Export in several formats
AgVerdict, AgVerdict, USA	Desktop and mobile app. Enable data delivery to regulatory agencies or packers, shippers, and processors. Data security, decision-making, variable-rate application possibility, soil analysis, and crop recommendations
Akkerweb, several companies from the Netherlands	Independent consulting platform for organizing field and crop rotation plans. Information in one central geo-platform
APEX/JDLink/My Operations/My Jobs, John Deere, USA	Online tools enabling access to farms, machines, and agronomic data. Allows collaborative decisions from the same set of information to optimize logistics, plans, and direct in-field work
Advanced Farming Systems (AFS), CASE IH, USA	Single, integrated software package. View, edit, manage, analyze, and utilize precision farming data to generate yield or variable-rate prescription maps. Maps and reports can be shared in different formats
Connected Farm/Farm Core, Trimble Agriculture, USA	Input, access, share records (images, reports) in real time. Integrates the whole system: crop scouting, grid sampling, fleet management, and contracts. *Farm Core* connects all aspects of the farm operation
Cropio, New Science Technologies, USA	Productivity management system. Remote monitoring of land. Real-time updates on current field and crop conditions; harvest forecasting. Web-based service and mobile app
Cropwin Vintel, ITK, France	Customizable tool for integrated crop management. Observation, analysis, and optimization. Decision support tool for vineyards. Tracks water status, cover crop, and nutrient management

(continued)

Table 15.1 (continued)

Platform name, company, and country	Relevant features
The Phytech Platform, PHYTECH, Israel	Plant-based app for irrigation. Monitors and provides data on crop growth. All data can be used to determine overall water needs
ESE Agri Solution, Source Trace, USA	Thought to manage group of farms and farmers. Unified and up-to-date farmer database. Record field visits with photos, notes, activities, and location. Farm-to-fork traceability of produce
Farmbrite, Farmbrite, USA	Farm schedule at-a-glance or in detail. The schedule can be shared to set up daily or recurring tasks. Weather forecast available. To-do list, reminders, events, and appointments
FarmCommand, Farm Edge, Canada	Farm management platform. Provides both hardware (weather station) and software for in-field decision support. Available as a web-based tool and a mobile app
Farmleap, Farmleap, France	Comparison of field performance locally and nationally. Reports time spent by operation type, yield analysis, production costs, irrigation follow-up, detailed weather, data sharing, employee management
FarmLogic/FarmPAD, TapLogic, USA	Web-based record-keeping. Global Navigation Satellite System (GNNS) field mapping to draw boundaries, mark points, measurements, etc.; personalized reports for distribution, pesticide database, maintenance records, and work orders creation
Farm Management Pro, Smart Farm Software, Ireland	Mobile app for farm records, costs and expenditure accounting, tractor management, crop management, fertilizer and spray compliance, staff timesheets, document management
Farmplan, Proagrica, United Kingdom	For crops, livestock, and business. Exchange data, work plans setup, weather data, data storage, instantaneous reports, and pesticide information
FieldView, The Climate Corporation, USA	Data connectivity and visualization, crop performance analysis, field health imagery. Offers variable-rate prescriptions and fertility management based on models
Granular, DowDuPont, USA	Different software according to necessities. Combination of several sources to build decision-making models. Advisory and training services. Support for more than 230 crop subspecies. Cloud-based
KSAS, Kubota, Japan	Cloud-based agricultural management support service integrated by Kubota machinery. For smartphones and PC. Farm management by collecting and utilizing data from supported machinery
Mapgrower, Agropreciso, Chile	Company-oriented platform that allows automated planning, work management, traceability, online statistics, account management, or visualization on maps. Available for smartphones
Myeasyfarm, Myeasyfarm, France	Allows to define fields and their operations, plan season work and share it with a team, see real-time progress, and analyze results
My Farm Manager, My Farm Manager, Canada	Packages available for variable-rate application, agronomy, and soil testing. Advice from experts. Marketing plans. Inventory and scheduled tasks in Croptivity application
Phoenix, Agdatam, Australia	It is modular so farmers can build their solutions. Available in the cloud or desktop. Farmers can create maps (.shp, gpx, pdf, bmp, and jpg formats), add data, and update them

(continued)

Table 15.1 (continued)

Platform name, company, and country	Relevant features
Precision Land Management (PLM) Connect, New Holland, Italy	Enables connection with field machinery. Map and analysis of crop/soil data, yield performance, variable-rate prescription, inventory and accounting records on supplies, seeds, chemicals, and fertilizer
SST software, Proagrica, United Kingdom	Collect and manage data in the field. Statistical analysis reports, decision-making tools
SMS, AgLeader, USA	Soil sampling, grids, and regions. Seeds with higher yield potential can be chosen based on historic performance, reports, record operations, variable-rate application maps, and prescriptions. Mobile app available
SpiderWeb GIS, Agrisat Iberia, Spain	Allows consultation, management, and analysis. Satellite images and other spatial reference layers. Data corresponding to each pixel can be downloaded in the form of temporary tables and graphs
Telematics, Claas, German	Collects important operational data for a self-propelled harvester and transfers it to a web platform
TAP TM, Topcon, Japan	Topcon and other companies' equipment compatible. Traceability and connectivity. Data management for farmers, data analysis for agronomists, multi-user data management, and cloud-based data management
Visual Green, Visual NaCert, Spain	Web platform to store farmers' data. GreenStar and MyJohnDeere compatibility, costs control, agroclimate data, official field notebook (compulsory in Spain), and authorized products
Visual NaCert, ProGIS Software, Austria	GIS: raster/vector maps, *krigging*, import/export in DXF or shp, fast Sentinel images. With its own development environment (SDK), allowing programmers to link their database apps to maps

Source: Saiz-Rubio and Rovira-Más (2020)

to monitor crop health. It makes it possible to generate maps of fixed-rate or variable-rate localized applications and create management zones.

The Otmis brand was created by the Brazilian company Jacto to provide farmers with their technologies for precision agriculture and digital agriculture. Among some Otmis products related to the displacement of agricultural machinery in the field are the Light Bar, the Omni 700 Electric Pilot, and the Omni 700 Hydraulic Pilot. These devices allow automatic routing and precision of machines in operations from planting to harvesting. Otmis has the OtmisNET telemetry system, a digital platform that allows the farmer to monitor agricultural operations carried out with the sprayers and fertilizer distributors of the Jacto brand. Its main functions are: (i) location of machines and their working path and (ii) reports such as applied volume maps, overlapping, speed, time ranges, ambient temperature, downtime reasons, relative humidity, fuel consumption, engine temperature, and spraying quality.

15.4 Data Collection with UAV

This topic presents the computational tools used to control UAVs and process the data obtained by sensors and cameras.

15.4.1 Drone Deploy

Drone Deploy is a UAV flight planning platform for mobile devices with Android and iOS operating systems and computers. After image acquisition, the cloud computing service of the Drone Deploy platform allows the generation of mosaic and processing of images, generating information such as vegetation indices, elevation, plant count, and a three-dimensional model of the terrain.

15.4.2 Agisoft – MetaShape

Agisoft (Fig. 15.4) is a computer program dedicated to the processing of aerial images obtained by a UAV, through photogrammetry. With this program it is possible to perform the topographic survey and generate high-quality 2D and 3D images of objects and locations, being widely used in the construction and agriculture sectors.

Fig. 15.4 Agisoft Metashape – Three-dimensional representation from UAV images. (Source: Agisoft 2020)

15.4.3 Pix4D

Pix4D is a Swiss company that develops software and applications focused on photogrammetry and computer vision, capable of processing multispectral images. Among the software programs developed by Pix4D, Pix4Dmapper and Pix4Dfields stand out in digital agriculture.

Pix4Dmapper is a photogrammetry software program that uses images captured by UAVs. The program allows the generation of georeferenced and elevation digital maps; point cloud; digital surface and terrain models; orthomosaics; as well as the calculation of distances, areas, and volumes. This software program has been employed in sectors such as construction, mining, public safety, and agriculture. Pix4Dfields features functionality like those of Pix4Dmapper, besides generating maps of vegetation indices, management zones, and prescription maps for variable-rate application. Both programs allow integration with different platforms and programs of the GIS and CAD (Computer Aided Design), which helps in the migration from one system to another, having a simple and user-friendly interface.

15.4.4 Sensix

The company's platform makes it possible to analyze and understand the spatial variability of plots from UAV and satellite images. The collected data are sent to the cloud, where they are processed. The results generated through this processing are application maps, planting rows and failures, zones of soil classification, plant population, biomass, chlorophyll content, water stress, and weed detection.

15.4.5 Taranis and TerrAvion

The Israeli company Taranis operates in the segment of analysis of aerial images (UAV, airplanes, and satellite) for decision-making. The UHR (Ultra-High Resolution) technology developed by the company makes it possible to generate images with resolutions between 8 and 12 cm/pixel. These images are processed with artificial intelligence technologies and used for the detection of pests, diseases, and weeds.

TerrAvion offers aerial image collection and processing services, operating in North America (USA and Canada) and South America (Brazil and Paraguay). The results generated can be imported and analyzed in the main digital agriculture platforms. The images are collected in multiple spectral bands (Table 15.2), allowing the calculation of vegetation indices such as NDVI and CIR.

15.5   Mobile Applications

Public sector institutions, agencies, and companies in many countries have developed various mobile applications and distributed them to the public. For example, the Brazilian Agricultural Research Corporation (EMBRAPA) has developed mobile applications and made them available. Some of these applications were developed with small- and medium-sized producers as target audiences. The applications made available by EMBRAPA in the app store for mobile devices with the Android operating system, specific for agriculture, are presented (Table 15.3).

15.6   Challenges and Future of Tools for Digital Agriculture

The main challenge to the advancement of digital agriculture in many countries is related to the unavailability of the Internet on farms. For example, according to the 2017 Agricultural Census (IBGE 2019), of the 5.07 million Brazilian rural establishments, 3.64 million (corresponding to 71.8%) do not yet have access to the Internet. However, the efforts of companies in the sector, governments, and farmers themselves tend to change this scenario in the coming years. Another challenge is the lack of trained professionals for the operation and maintenance of the technologies presented in this chapter. In this context, educational institutions with technical and higher education courses play a fundamental role in the training and professionalization of people.

Overcoming these challenges in the coming years is essential, as digital agriculture tools have great potential to make the agricultural production system more efficient. Among the demands of the agricultural scenario, Digital Agriculture directly contributes to the increase of food production and greater sustainability of agricultural production.

There is a tendency to the development of computational tools with high levels of autonomy, especially in decision-making, that is, with minimal human

Table 15.2 Spectral bands collected by TerrAvion

Spectral bands	Wavelength (nm)
Blue	430–500
Greed	490–580
Red	575–700
Near-infrared	830–800
Green 2	545–590
Red 2	576–652
Thermal	7500–13,000

Table 15.3 Application developed by EMBRAPA and made available in the app store for Android devices

Name	Functions
Zarc – Plantio Certo	Presents the best planting dates for 43 crops, based on data from the Agricultural Zoning of Climatic Risk (*Zoneamento Agrícola de Risco Climático* – ZARC). Quantifies the risks arising from adverse climatic conditions and allows each municipality to identify the best time for planting the crops, in the different soil types and crop cycles, according to the characteristics and needs of each crop
Bioinsumos	National Catalog of Bio-inputs to facilitate and promote access to information about products available for use and where to find them
Nutrisolo	Recommendation of soil fertilization and liming for pineapple, banana, citrus, and cassava crops in the Amazon. The application also provides some calculation tools and practical tips on the management of these crops
AgroPragas Maracujá	Information on the main pests and diseases of passion fruit crop and their forms of control
Doutor Milho	Identifies the development stage of maize plants and offers a series of management suggestions, specific for each stage. Consultation on all maize cultivars available on the market in each season
Guia Clima	Agroclimatic monitoring system that provides, in real time, data on weather conditions (temperature, relative humidity); information (averages, normal); and alerts (low relative humidity, strong winds, frosts)
Orçamento Forrageiro	Quantifies the biomass of forages of native pastures of the Caatinga biome. The application simulates the balance between the supply and the need for forage (according to the number of animals registered), providing a report that informs how much of the demand is being met by the pasture
Pastejando	Collection of forage species for consultation
FertOnline	Fertilizer recommendation for coconut (dwarf and giant), maize, and orange crops, based on soil and leaf analyses. The data that support recommendations for the crops were obtained through long-term field experiments, whose results were published in the form of articles
Agritempo	Access, via the Internet, to meteorological and agrometeorological information from several Brazilian municipalities and states. In addition to informing the current climate situation, the system's database supports the development of the recommendations of the Agricultural Zoning of Climatic Risks (ZARC), a policy maintained by the Brazilian Ministry of Agriculture, Livestock and Food Supply
Guia InNat	Consultation of images of natural agents of agricultural pest control, information on predators and parasitoids, morphological characteristics of the main families, as well as life cycle and function of these arthropods as pest controllers. This application also makes it possible to compare an insect photo taken with the cell phone camera with the images from the Guide gallery for recognizing the natural enemies of agricultural pests
Ferti-Matte	Helps agrarian science professionals to interpret and recommend the fertilization of herbs in the stages of planting, canopy formation, and production
Manejo-Matte	Assists in the diagnosis of commercial herbs and suggests improvements in management based on technologies to produce yerba mate recommended by EMBRAPA

(continued)

Table 15.3 (continued)

Name	Functions
PlanejArroz	Access to a set of information from 131 municipalities that produce irrigated rice in Rio Grande do Sul. It indicates when each of the six most important stages of plant development should occur, for 41 cultivars, through the average dates (30 years) and the crop season year, and the respective deviations between them. This information is very important for planning and decision-making about crop management. In addition, PlanejArroz also allows the estimation of yield, in the average of years and in the crop season, and the deviation between these two variables, of the three most sown cultivars in the state
BioSemeie	Management of seed stocks and systematization of the seed catalog with their morphological and adaptive characteristics to contribute to the conservation of agrobiodiversity of heirloom seeds, as well as in research related to the genetic improvement of these species
Monitora Oeste	Pest and disease monitoring system for Western Bahia
Nutri Meio-Norte	Evaluation and leaf diagnosis using nutritional balance tools. The first version was conceived with the creation of a database of crops representative of the Mid-North Region with the analysis of plant tissue from the diagnostic leaf (macro- and micronutrients) and the survey of the yield of plots of commercial areas cultivated with soybean (*Glicine max*), in the states of Piauí and Maranhão

Source: Google Play (2020)

intervention. From these platforms and communication between machines, it is possible that data are collected, decision-making is performed, and operations are carried out without any human intervention. For this, the technologies presented in other chapters, such as the Internet of Things, processing of large data sets (big data), use of machine-learning algorithms, and artificial intelligence, are essential.

Aimed at gathering several data in one place, facilitating analysis and decision-making, the diffusion of API (Application Programming Interface) on the platforms presented in this chapter has been observed. Through the API, one platform can communicate with another, even if it belongs to another developer. For example, the Leaf Agriculture platform allows connection with platforms of companies such as Trimble, Climate Fieldview, CNH, John Deere, AGCO, AgLeader, Stara, and Solinftec.

Abbreviations/Definitions

- Application Programming Interface (API): is a feature that allows the interaction of software (or platforms) with others.
- Agricultural Zoning of Climatic Risks (ZARC): is the delimitation of regions that are considered suitable for the development of an agricultural crop, taking into consideration climate conditions, as well as the thermal and hydrological demands of the species.
- Brazilian Agricultural Research Corporation (EMBRAPA): is a state-owned research company, affiliated with the Ministry of Agriculture of Brazil, with the mission of developing research, development, and innovation solutions for the sustainability of agriculture, for the benefit of Brazilian society.

- Color Infrared (CIR) image: also called false-color image, is an image composed of near-infrared (NIR), red, and green bands of electromagnetic spectrum.
- ISOBUS standart: is standardization of communication between tractors and machines attached, defined by the ISO 11783 standard, allowing the sharing of information between them without any restrictions.
- Farm Management Information Systems (FMISs): are computational tools with the function of assisting the farmer in various tasks, from planning and execution of operations to management of data and documents.
- Geographic Information System (GIS): is a set of integrated software, hardware, data, and users that make it possible to collect, store, manage, and analyze geo-referenced data, enabling the production of information from their use.
- Global Navigation Satellite System (GNNS): is a system that uses satellites to provide geo-spatial positioning.
- Normalized Difference Vegetation Index (NDVI): is an index used to quantify vegetation by measuring the energy reflected in the red and near-infrared (NIR) bands of the electromagnetic spectrum.
- Telemetry: is a digital technology that allows the sending of data acquired by sensors to a storage server ("the cloud") using the Internet.
- Unmanned Aerial Vehicles (UAV): is an aircraft without any human pilot, crew, or passengers on board, operated by remote control.
- UHR (Ultra-High Resolution) image: technology that allows obtaining images with very high spatial resolution.

Take Home Message/Key Points

- Farm Management Information Systems assists the farmer in various farm activities.
- Various Farm Management Information Systems of global dissemination or dissemination in a country or region.
- Computational tools for unmanned aerial control and process data obtained by sensors connected to it.
- Mobile applications developed by public institutions, agencies, and companies for use in agriculture.

References

Agisoft (2020) Tutorial (beginner level): orthomosaic and DEM generation with Agisoft PhotoScan Pro 1.2 (without ground control points). Available in https://www.agisoft.com/pdf/PS_1.2% 20-Tutorial%20(BL)%20-%20Orthophoto,%20DEM%20(without%20GCPs).pdf. Access in Nov of 2020

AgLeader (2020) What is ISOBUS? Available in https://www.agleader.com/blog/what-is-isobus-2/. Access in Nov of 2020

Brazilian Institute of Geography and Statistics – IBGE (2019) Agricultural census 2017. Definitive results. Rio de Janeiro 8:1–105. (In Portuguese)

Kverneland Group. ISOBUS plug to tractor. Availabe in: https://ien.kverneland.com/Images-for-iM-FARMING-articles/ISOBUS-plug-to-tractor Access in June of 2022.

Google Play (2020) Apps. Available in https://play.google.com/store/search?q=embrapa&c=apps. Access in Nov of 2020

Saiz-Rubio V, Rovira-Más F (2020) From smart farming towards agriculture 5.0: a review on crop data management. Agronomy 10:207–228

Chapter 16
Digital Data: Cycle, Standardization, Quality, Sharing, and Security

Antonio Mauro Saraiva, Wilian França Costa, Fernando Xavier, Bruno de Carvalho Albertini, Roberto Augusto Castellanos Pfeifer, Marcos Antonio Simplício Júnior, and Allan Koch Veiga

16.1 Introduction

The main characteristic of what is called Digital Agriculture, or Agriculture 4.0, is the intensive use of data. It can be said that Digital Agriculture is data-driven. In other words, data, which are becoming increasingly available with spatial and temporal attributes, at high frequencies and on an unprecedented scale, have become essential inputs for the processes that culminate in decision-making.

This digitization phenomenon that is now occurring in agriculture repeats what has already occurred in other areas of human activity, which have been more agile in the incorporation of Information and Communication Technologies (ICTs) in their processes. In fact, the agricultural sector as a whole was slow to adopt these technologies compared to many other sectors and is still one of the least digitized in the world (Krishnan 2017). However, this situation is changing rapidly and, of course, varies greatly among different countries and different sectors of agriculture, including animal production.

It is worth noting that Digital Agriculture has not appeared suddenly, and what we currently see is the result of a long process that began when the first analog electrical monitoring and control systems were incorporated into agricultural tractors and facilities, in the early twentieth century. It gained great momentum with the development of electronics, microelectronics, and ICTs, which occurred from the second half of the twentieth century, as reported by Cox (1997).

A. M. Saraiva (✉) · W. F. Costa · F. Xavier · B. de Carvalho Albertini ·
M. A. Simplício Júnior · A. K. Veiga
Universidade de São Paulo, Escola Politécnica, São Paulo, Brazil
e-mail: saraiva@usp.br; fxavier@usp.br; balbertini@usp.br; msimplicio@usp.br

R. A. C. Pfeifer
Universidade de São Paulo, Faculdade de Direito, São Paulo, Brazil
e-mail: roberto.pfeiffer@usp.br

© The Author(s), under exclusive license to Springer Nature Switzerland AG 2022
D. Marçal de Queiroz et al. (eds.), *Digital Agriculture*,
https://doi.org/10.1007/978-3-031-14533-9_16

From the 1990s, Precision Agriculture (PA) developed remarkably based on the technology already available and was certainly one of the milestones for the transition to Digital Agriculture. This is because PA demands intense use of data acquisition systems in the field, control systems in the machines for the application of inputs at variable rates, and information systems to process an unprecedented amount of data, in spatial and temporal scales. PA boosted digitization, as an important market began to develop and the technology – used widely in other economic sectors – was mature and with more accessible costs for adoption in the field.

After that and again following what was seen in other sectors and industries, there was a very large growth in the development of equipment and systems for data collection, automatic control, information management, and support for decision-making. More recently, a multitude of applications for mobile platforms, smartphones, and tablets to support the most diverse agribusiness activities have also appeared.

Many data-related issues are essentially dependent on the context in which they are inserted and in which they are used. It is thus impossible to detail all those issues for each agricultural use in the space of this text. However, it is important to draw attention to more general aspects that must be considered in any application. In this chapter, some of these aspects are addressed: data life cycle and data science, data standardization, data quality, and data security and legal aspects arising from new data protection laws, recently approved in many countries.

16.2   Big Data, Data Science, and Data Lifecycle

Data is characterized as the "oil" of the digital age.[1] This association is not new: the expression "data is the new oil," credited to the English mathematician Clive Humby, in 2006, has been widely used to characterize the importance of data in the era of Big Data. This term is used to define the high volume of data, whose collection, storage, circulation, and sharing require specific technology and analytical methods for their transformation into value by companies (EC 2015; Boyd and Crawford 2011).

Traditional approaches to data analysis are not adequate enough in the Big Data era. In addition to the large volume of data, the diversity in the formats and sources and the speed at which they are generated require an evolution in the techniques, methods, and tools. Thus, besides traditional statistical methods, there is a need to use methods from Computer Science to collect, transform, integrate, and analyze data. Data Science meets this demand and is characterized by a multidisciplinary approach in which, in addition to professionals in Statistics and Computer Science, other areas are necessary. In the case of Big Data in agriculture, professionals from the application area are fundamental in projects to extract useful information from agricultural data.

---

[1] The Economist. Available at: https://www.economist.com/leaders/2017/05/06/the-worlds-most-valuable-resource-is-no-longer-oil-but-data

Artificial Intelligence (AI) methods have gained even more importance in the Big Data scenario, both in the search for greater efficiency and for revealing information that would not be obvious only with the application of traditional data analysis methods. However, data analysis goes far beyond the application of these methods and AI, being a part of the data life cycle that contains different activities, from planning to analyzing and producing results, as illustrated in Fig. 16.1.

This data cycle illustrates, generally speaking, data management activities, both in a scientific context and a business context. Although each stage of this cycle can be performed by a different professional, the importance of the multidisciplinary vision brought by Data Science is highlighted.

Experts in the application domain, in this case, experts in agriculture (in a broad sense), have a fundamental role from the very planning stage, in which the objectives in the use of these data will be defined. Therefore, before defining what data to collect, it is necessary to establish the objectives and sources. However, the participation of domain experts is not limited to the planning stage. In fact, it can be and often is also related to other activities in the cycle defined in Fig. 16.1. This reinforces the importance that professionals linked to the agricultural sector – agronomists, zootechnicians, agricultural engineers, agricultural technicians, etc. – increasingly have training that allows them to interact with digital technology professionals.

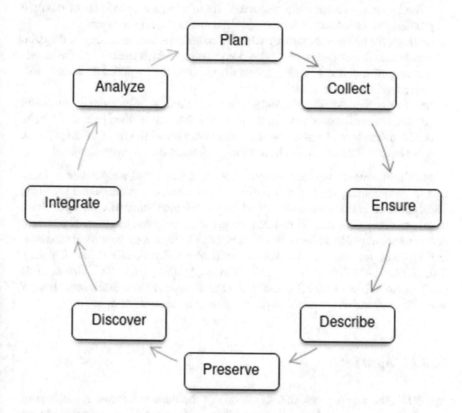

Fig. 16.1 Data life cycle model. (Source: Adapted from Silva 2017)

16.3 Standardization of Data and Communication in Agriculture

Agricultural data are quite diverse from the point of view of their sources and formats and do not usually follow a widely accepted standardization. This stems from the very characteristic of the sector, which is geographically very dispersed and technologically very uneven. Data standardization is one of the key factors for the success of digital agriculture. It allows two entities (software, people, institutions, etc.) to exchange data that will be interpreted and treated in the same way, regardless of physical or temporal distance, avoiding errors and reducing the costs related to data conversion.

Data standardization is a set of collaborative documents that indicate the consensus of a specific community on the representation, format, definition of meaning, structuring, marking, transmission, manipulation, use, and management of data. Below are the main benefits of standardization:

- It is described by qualified people. When an entity sees the need for standardization or when the volume of data on a particular subject increases significantly, interested people are brought together in public calls, in which the definitions behind a standardization are discussed. These groups are made up of multiple profiles, from academic, business, government, and local producers.
- It allows for better transparency and a homogeneous understanding of the data. Standardizations often have a related vocabulary, which means that the understanding of a concept is also standardized, that is, two different entities will understand the data in a similar way.
- Saves resources. Although the adoption of standards is often more expensive in relation to non-standard data, in the long run, that cost is often paid since tools, codes, methods, and resources can be reused without the need to adapt them. Standardized data are said to have a longer lifetime than non-standardized data.

Most standards are written and formalized by entities with a reputation or mandate in certain areas, such as associations, governments, or professional societies. Through these entities, some data and communication standards for agriculture have been developed, being at different stages of maturity and adoption. If you have questions about which standard you should use, please contact your local standardization body, for example, ARSO (Africa), ABNT/Brazil, SCC/CSA (Canada), SAC (China), AFNOR (France), DIN (Germany), UNI (Italy), JISC (Japan), BSI (UK), or ANSI (USA). ISO (International Organization for Standardization, https://www.iso.org/) can be contacted to point out your local standardization body.

16.3.1 AgroXML

AgroXML (http://www.agroxml.de/) is one of the main standards in the field of agriculture since it covers a wide range of topics, from precision agriculture to the food production chain or the management of smart agricultural companies. It is based on the XML standard (eXtensible Markup Language), whose main features are ease of use and extensive support. The standard was developed independently by a study group at the University of Hohenheim (Stuttgart, Germany) in 2004. However, the flexibility resulting from the choice of XML has spread the standard worldwide, which is one of the main data standards used. Note that AgroXML is open-source, and its use is free.

16.3.2   AgMES

The AgMES (http://aims.fao.org/standards/agmes/, Agricultural Metadata Element Set) was developed to organize information from the agricultural area, including any digital information. Today, it is maintained by the Agricultural Information Management Standards (http://aims.fao.org/) of the United Nations Food and Agriculture Organization (FAO), but its development is stagnant. New adoptions of the standard are discouraged, but it is still useful for organizers of information (e.g., libraries, collections, etc.). For new adoptions, it is suggested to use AGRIS.

16.3.3   AGRIS

AGRIS (http://aims.fao.org/agris-network), also maintained by FAO, is the successor to AgMES. It aims to catalog information on food production in general, with a focus on agriculture (including precision agriculture). The standard is predominantly bibliographic, with significant use in academia, but little commercial adoption, which is why we will not extend its description.

16.3.4   AGROVOC

AGROVOC (http://aims.fao.org/standards/agrovoc/) is another mechanism maintained by FAO, consisting of extensive multilingual vocabulary (including English and Portuguese) on all aspects of food production. It is used as a guide for data storage and communication. Using AGROVOC, data interpretation should be clear and unambiguous, as those involved have a unique source of concepts, terms, and relationships between them. The standard covers about 36,000 concepts, which makes it the main semantic reference for agricultural standards. Its use is free through a Creative Commons license.

16.3.5   ISOBUS

ISOBUS is the main standard in the agricultural area for use in machinery. It is a communication standard used for data interoperability of machines and implements (M2M). Its adoption allows a producer to purchase a machine from one manufacturer and service or implement from another, for example, as long as both are adherent to the standard.

This standard consists of a high-resilience serial communication based on CANBus, the communication network used in non-agricultural vehicles, such as automobiles and trucks. It involves the communication protocol (physical and logical data format), the user interface in the machines (e.g., the on-board computer on the tractor), control of operations, and file servers. It also has a linked vocabulary, which allows data collected through ISOBUS to be interpreted in the same way by tools from different manufacturers. The main advantage of ISOBUS is the economy, since, by adopting it, the producer avoids redundancy in the equipment and can reuse or connect machines and implements from different suppliers. However, as message exchanges and file formats are also standardized, the standard has also been used in data analysis. It is an international standard (ISO 11783) broadly adopted by the agricultural machinery industry worldwide. It is strongly recommended that any implement or machine purchased be adherent to the standard.

16.3.6   Open Geospatial Consortium Standards (OGC®)

We cannot discuss data standardization in agriculture without referring to the standards specified by the Open Geospatial Consortium (OGC). The OGC is an international consortium of more than 530 companies, government agencies, research organizations, and universities geared toward making geospatial information and services (localization) Findable, Accessible, Interoperable, and Reusable (the so-called FAIR principles) (Wilkinson et al. 2016). Created in 1994, the consortium was aimed at standardizing data used in Geographic Information Systems (GIS). Today, OGC operates in the development and implementation of open standards for geospatial content and services, sensor networks, the Internet of Things, georeferenced data processing, and data sharing (Mckee 2020).

Several OGC standards have become popular over the years, as the use of geospatial data evolved. Web service standards, such as the Web Map Service ((WMS) vector map bitmap rendering service), Web Feature Service (service for accessing and editing geometries (vector data)), and the Web Coverage Service (service for data rendering raster), are currently supported on all GIS tools and servers in the market. Map servers, such as Geoserver (http://geoserver.org/), MapServer (https://www.mapserver.org/), and the adoption of these standards in GIS tools already consolidated in the market, such as ESRI ArqGIS Server (https://www.esri.com/en-us/

arcgis/products/arcgis-enterprise/overview), demonstrate the success in using these standards, in different communities and with different focuses.

Currently, OGC has working groups directly concerned with the development of specific standards related to agricultural resources. We here highlight the **Agriculture Domain Working Group** (DWG), some purposes of which are (Di and Charvat 2020): to examine and propose the possibilities of aligning and harmonizing agricultural information exchange standards between initiatives and organizations, such as CEFACT (UN), ISO TC 23, ISOBus, AgroXML, OGC, W3C, etc.; and the development of a reference architecture for the use of coding standards and the OGC interface in common agricultural activities.

16.4   AgGateway Standards

AgGateway is a global nonprofit organization, whose members develop standards and other resources so that agricultural companies can access information quickly by adopting standards for interoperability, facilitating the transition to digital and sustainable agriculture (https://www.aggateway.org/AboutUs/Mission.aspx). The Ag Data Application Programming Toolkit (ADAPT) consists of an Agricultural Application Data Model, a common API (Application Programming Interface), and a combination of proprietary and open-source data conversion plug-ins. Companies that market Agricultural Management Information Systems are responsible for building their own implementation of mapping the Agricultural Application Data Model to their specific data model. They include several standards (Ferreyra 2017), such as for irrigation data (PAIL – Precision Ag Irrigation Language) (Aggateway 2020a), the integration of planting data and the use of fertilizers (SPADE – Fertilization Data Standards) (Aggateway 2020b), and semantic image identification for agricultural remote sensing in GeoTIFF format (PICS – Imagery Tagging) (Ferreyra 2019).

16.5   Data Quality

Data quality has significant consequences and effects on its use. Poor data quality is estimated to cause 8–12% of revenue loss and to represent 40–60% of companies' service costs (Redman 1998).

In the agricultural sector, there are great benefits in assuring the quality of the data used by specialists for decision-making and for supporting data-dependent activities, such as income forecasting, monitoring, and planning (Malaverri and Medeiros 2012). Therefore, the assessment and management of data quality for the improvement of "data fitness-for-use" are actions justified by cost reduction and by making more assertive and efficient decisions.

16.6   Quality Assessment

In general, management methodologies presuppose clear and objective definitions of success metrics. In the field of data quality, it is not different. Improving this quality depends on clear metrics of success, the so-called data quality dimensions. Therefore, there is a consensus in the literature that data quality is a multidimensional concept, and its meaning in a project is defined by a series of relevant dimensions in a given context, such as consistency, precision, accuracy, completeness, trust, reputation, accessibility, among others (Wang and Strong 1996).

Identifying which data quality dimensions are relevant to the success of a project, measuring the quality in these dimensions, and establishing criteria to assess whether that quality is fit-for-use in a given context are essential for efficient data quality management (Veiga et al. 2017).

16.7   Quality Management

The objective of this management is to improve the quality of the data by the prevention and correction of errors that directly or indirectly degrade the quality of the data in one or more relevant dimensions of a project.

Two strategies can be adopted for this: quality control and quality assurance. The first seeks to optimize the measures of the data quality dimensions whenever possible, without losing data but assuming that the quality may not be fully in accordance with the design criteria. Conversely, quality assurance assumes that the data have a quality level totally in accordance with the project criteria, which may imply data loss if a subset of data does not meet the pre-established criteria (Veiga et al. 2017).

Regardless of the strategy adopted over time in the organization, the data used to support punctual decision-making must be evaluated with quality measures (e.g., precision, accuracy, consistency, completeness), according to the pre-established criteria in the context of the project.

16.8   Data Sharing and Security

Many companies have been increasing their profits by using information about users' behavior, preferences, needs, expectations, desires, and opinions. Several of the innovative businesses from the digital age rely on user data that are quite often shared among different platforms. For this reason, many already consider data as a commodity (Morando et al. 2014). Data collection, analysis, and customization are also part of agribusiness, whereby it is used to improve products, increase sales, or learn about consumer preferences and adapt to them (through advertising). For this reason, some authors highlight the existence of a data value chain (Curry 2016; Miller and Mork 2013) in agribusiness.

In this scenario where data sharing takes a major role, it is important to analyze which security aspects are relevant and which actions should be taken to ensure compliance with data protection laws, such as the European General Data Protection Regulation (GDPR) or the Brazilian General Data Protection Law (*Lei Geral de Proteção de Dados* – LGPD).

16.9   Data Security

When allowing data to be shared among different institutions, it is worth considering some essential aspects of information security. Typically, a robust system should provide a combination of the following basic security pillars:

- *Confidentiality or secrecy* – Only authorized users can access information transmitted or stored in the system. For example, personal data of customers and suppliers should never be transferred in cleartext over the Internet, but instead use secure communication technologies (e.g., in the case of a web page, HTTPS should be used instead of HTTP). This approach aims not only at protecting the privacy of any entity whose data is shared but also to avoid any possible competitive damage resulting from strategic information leakages.
- *Integrity* – If some piece of information is modified without authorization, whether accidentally or on purpose, it must be possible to detect this modification. Note that it is not always possible to prevent or undo the change, but it is essential to enable detection aiming to prevent misguided actions, taken after the analysis of bogus data. For example, when using sensors to continuously monitor soil quality, the integrity of the collected data must be protected against modification; otherwise, this might lead to the application of an undue amount of fertilizers, possibly compromising the entire crop.
- *Authenticity* – During the whole duration of a communication, both sender and receiver should be able to identify each other's messages. This service is particularly important to prevent intrusion attempts, such as an attacker trying to impersonate a legitimate user of the system, in an attempt to access sensitive data; the insertion of malicious sensor nodes in a field monitored by an IoT system, aiming to inject false information into the system and, as a result, manipulate its actions.
- *Non-repudiation* – Guarantee that a user cannot deny having created or sent a message. This security service is directly related to the concept of digital signatures: when signing a document, one cannot subsequently deny such a signature, so the document author can be held responsible in case of misconduct. For this reason, the deployment of non-repudiation services into a system is usually a requirement for constructing robust audit mechanisms.

- *Availability* – Legitimate users of the system should not be prevented from accessing it. A common example of an attack against the availability of systems is the so-called Denial of Service (DoS) attack. Several mechanisms can be used to mitigate or lessen the impact of such threats. An example is to filter messages sent by automatic means, for example, by requiring the solution of a challenge that is not easily solved by robots (a mechanism commonly known as "CAPTCHA"). When it is necessary to support such automation, it is common to require senders to be authenticated, and then limit their transmission rate to avoid attempts to monopolize the system resources. Another common approach for dealing with the threat of overload involves the adoption of some degree of system redundancy and elasticity, which is commonly achieved by employing cloud computing technologies.

To identify which security services are a priority in each application scenario, it is important to consider the characteristics of the system and of the data it handles. For example, consider an automated irrigation system in which soil moisture is constantly monitored and sprinklers are activated whenever necessary. In this case, data integrity, authenticity, and availability are likely more important than confidentiality and non-repudiation; after all, avoiding excess or lack of water is more relevant than preventing third parties from accessing moisture-related readings, or proving that a given sensor was responsible for sending a specific reading. However, when the data transferred involves product prices, the result of negotiations, the contents of contracts, or other strategic information, all of the hereby listed security services can become similarly relevant.

Finally, it is worth noting that there are technologies created specifically to facilitate the task of securely sharing data among different organizations, taking into account the aforementioned security concerns. One particularly prominent solution is OAuth, or Open Authorization (https://oauth.net/2/), an open security protocol that enables data owners to delegate access to (part of) their data to a third party, for a specific time. The protocol is currently in version 2.0, and it is used in the construction of several solutions in which the control of the information flow is centered on users rather than on the servers employed for storing users' data.

16.10   Protection of Personal Data

Most countries exporting and importing agricultural products have personal data protection legislation. This is the example of Brazil and its Data Protection General Law, and the countries of the European Union whose General Data Protection Regulation, GDPR, was enacted in May 2016.

Thus, the collection and sharing of personal data of individuals involved in agricultural production must comply with the rules imposed by the applicable data protection legislation.

The best-known hypothesis of legal authorization is the consent of the data subject (Article 6, 1, of the GDPR). In this context, for example, if the agricultural producer wishes to collect information about the shopping habits of his consumers and share such data with third parties, he will have to obtain prior consent from the acquirers or, alternatively, anonymize such data.

Notwithstanding, there are other possibilities for using personal data that do not depend exclusively on the consent of the data subject.

Some examples can be given concerning the agribusiness sector. There are cases in which data collection and transmission to inspection bodies are mandatory (e.g., pesticide user data and the place where the product will be applied). In these cases, data processing is allowed regardless of the prior consent of the data subject (Article 6°, 1, c of the GDPR).

Another hypothesis occurs when the contract established between the consumer and the agricultural company requires certain data to be collected and shared. This would be the case, for example, of the farmer who buys seeds from a certain producer. There may be a need to collect and process the consumer's personal data to enable the contract conclusion. In this case, the collection and sharing of data between seller and producer do not require the prior consent of the data subject, because it is permitted by the contract.

However, clear and adequate information on the processing of the data collected will always be necessary, whatever the basis of the lawfulness of processing.

16.11 Final Considerations

Data is a new source of wealth, a precious asset for those who generate it and for everyone involved in the chain of its use, symbolized by the data life cycle. It is a fact already well known and explored in various industries and, currently, also in agriculture. Data is the basic input for obtaining information (i.e., contextualized data) and knowledge, which supports decision-making and the formulation of business and government policies.

However, there are countless aspects to be considered for the entire cycle to develop effectively in the best possible way, especially considering an area as complex as agriculture. This is an area where data is generated in a very distributed way, in time and space, by a huge variety of users, devices, equipment, and systems that are very heterogeneous from various points of view, including technological and cultural, among others.

In this chapter, some of these aspects were presented that apply in general since a specific approach for each use would not make sense in the scope of this text. With that, we tried to offer a first approach, which should be further explored by the reader. It is worth mentioning that, with the growing importance of data from current businesses, Data Science has emerged as an important area, which corroborates the fact that, in the space of a chapter, the possible approach is the introduction of the subject.

There is no doubt, however, that data is also one of the main assets of the agricultural sector and the main foundation of Digital Agriculture.

Abbreviations/Definitions
- AI: Artificial intelligence is the intelligence demonstrated by machines, as opposed to the natural intelligence displayed by animals including humans.

Take Home Message/Key Points
- The main characteristic of what is called Digital Agriculture or Agriculture 4.0 is the intensive use of data.
- Digital Agriculture is data-driven. Data, which are becoming increasingly available with spatial and temporal attributes, at high frequencies and on an unprecedented scale, have become essential inputs.
- Data must be considered in regard to its life cycle, standardization, quality, security, and legal aspects to be used in its full benefits to farming.

References

Aggateway (2020a) Precision Ag Irrigation Language (PAIL). Available at https://www.aggateway.org/eConnectivityActivities/Implementation/PrecisionAgIrrigationLanguage(PAIL).aspx. Accessed 12 Mar 2020

Aggateway (2020b) Specific Initiatives. Available at https://www.aggateway.org/GetConnected/SpecificInitiatives.aspx. Accessed 13 Mar 2020

Boyd D, Crawford K (2011) Six provocations for Big Data. In: Symposium on the dynamics of the internet and society: a decade in internet time, September. Available at https://ssrn.com/abstract=1926431 or https://doi.org/10.2139/ssrn.1926431. Accessed 13 Mar 2020

Cox SWR (1997) Measurement and control in agriculture. Blackwell Science, London. ISBN 0-632-04114-5

Curry E (2016) The big data value chain: definitions, concepts, and theoretical approaches. In: Cavanillas J, Curry E, Wahlster W (eds) New horizons for a data-driven economy. Springer, Chain

Di LP, Charvatdim F, Sakuda LO (org) (2020) Radar AgTech Brasil 2019: mapeamento das startups do setor agro brasileiro. Brasília: Embrapa, SP Ventures e Homo Ludens, K. Agriculture DWG. Available at https://www.ogc.org/projects/groups/agriculturedwg. Accessed 13 Mar 2020

EC – European Commission (2015) Towards a thriving data-driven economy. Off J Eur Union 58:61–65

Ferreyra R (2017) ADAPT Public Space. 15 set. Available at https://www.aggateway.org/GetConnected/AgGateway%E2%80%99sADAPT.aspx. Accessed 14 Mar 2020

Ferreyra, R (2019) A AgGateway Post-Image Collection Specification (PICS). Available at https://aggateway.atlassian.net/wiki/x/XABrDw. Accessed 13 Mar 2020

Krishnan N (2017) Cultivating Ag Tech: 5 trends shaping the future of agriculture. Available at https://www.cbinsights.com/research/agtech-startup-investor-funding-trends. Accessed 14 Mar 2020

Malaverri JEG, Medeiros CB (2012) Data quality in agriculture applications. In: GEOINFO, 13 November 25–27 de 2012, Campos do Jordão, SP, Brazil. Proceedings of the Brazilian Symposium on GeoInformatics. Campos do Jordão, SP, pp 128–139

Mckee L (2020) OGC history (detailed). Available at https://www.ogc.org/ogc/historylong. Accessed 13 Mar 2020

Miller HG, Mork P (2013) From data to decisions: a value chain for big data. In It Profess 15:57–59. https://doi.org/10.1109/MITP.2013.11

Morando F, Iemma R, Raitieri E (2014) Privacy evaluation what empirical research on users' valuation of personal data tells us. Internet Pol Rev 3(2). https://doi.org/10.14763/2014.2.283

Redman T (1998) The impact of poor data quality on the typical enterprise. Communications of the ACM. https://doi.org/10.1145/269012.269025

Silva DL (2017) Estratégia computacional para apoiar a reprodutibilidade e reuso de dados científicos baseado em metadados de proveniência. PhD thesis – Escola Politécnica, Universidade de São Paulo, São Paulo. https://doi.org/10.11606/T.3.2017.tde-05092017-095907. Available at https://www.teses.usp.br. Accessed 3 Jan 2020. (English title: Computational strategy to support reproducibility and reuse of scientific data based on provenance metadata)

Veiga AK, Saraiva AM, Chapman AD, Morris PJ, Gendreau C, Schiegel D, Robertson TJ (2017) A conceptual framework for quality assessment and management of biodiversity data. PLOS One 12:e0178731. Available at https://doi.org/10.1371/journal.pone.0178731. Accessed 3 Jan 2020

Wang RY, Strong DM (1996) Beyond accuracy: what data quality means to data consumers. J Manag Inf Syst (Published by M.E. Sharpe, Inc.) 12:5–33. Available at http://www.jstor.org/stable/40398176. Accessed 3 Jan 2020

Wilkinson MD, Dumontier M, Aalbersberg IJ, Appleton G, Axton M, Baak A, Blomberg N, Boiten J, da Silva Santos LB, Bourne PE, Bouwman J, Brookes AJ, Clark T, Crosas M, Dillo I, Dumon O, Edmunds S, Evelo CT, Finkers R, Gonzalez-Beltran A, Gray AJG, Groth P, Goble C, Grethe JS, Heringa J, Hoen Peter AC, Hooft R, Kuhn T, Kok R, Kok J, Lusher SJ, Martone ME, Mons A, Packer AL, Persson B, Rocca-Serra P, Roos M, Schaik R, Sansone SA, Schultes E, Sengstag T, Slater T, Strawn G, Swertz MA, Thompson M, Lei J, Mulligen E, Velterop J, Waagmeester A, Wittenburg P, Wolstencroft K, Zhao J, Mons B (2016) The FAIR guiding principles for scientific data management and stewardship. Sci Data 3:160018. Available at https://doi.org/10.1038/sdata.2016.18. Accessed 3 Jan 2020

Chapter 17
Case Study: SLC Agrícola

Ronei Sandri Sana

17.1 Introduction

SLC Agrícola, founded in 1977 by the SLC Group, is one of the world's largest producers of grains and fibers, and is focused on the production of cotton, soybeans, and corn. Headquartered in Porto Alegre (RS, Brazil), the company has 22 farms, strategically located in seven Brazilian states, totaling 660,000 ha with crop.

The business model is based on a modern production system, with high scale, standardization of farms, high technology, cost control, and social and environmental responsibility. Throughout its history, SLC Agrícola has developed solid expertise in prospecting and acquiring land in new agricultural frontiers. The process of acquiring areas with high productive potential also aims to capture the real estate value that arable land in Brazil provides due to its comparative advantages in relation to the main agricultural producers in the world, such as the United States, China, India, and Argentina.

SLC Agrícola believes that excellence in the management of economic, social, and environmental aspects results in the reduction of the environmental impacts of its operations. Sustainability in the company is based on these precepts and on the "Big Dream of the Company": positively impacting future generations, being the world leader in efficiency in the agricultural business and in respect for the planet. The programs and actions adopted at SLC Agrícola have economic management as components, as well as the adoption of best agronomic practices and technologies, to minimally impact the environment. In this sense, digital agriculture was an essential step for SLC to achieve these goals.

R. S. Sana (✉)
Master Precision Agriculture, Digital Agriculture Manager at SLC Agrícola,
Porto Alegre, Rio Grande do Sul, Brazil
e-mail: ronei.sana@slcagricola.com.br

© The Author(s), under exclusive license to Springer Nature
Switzerland AG 2022
D. Marçal de Queiroz et al. (eds.), *Digital Agriculture*,
https://doi.org/10.1007/978-3-031-14533-9_17

17.2   Digital Agriculture at SLC Agricola

For all sizes of rural properties, from small to large agricultural companies, there are technologies that can increase production efficiency. Digital Agriculture makes it possible to solve common problems faced by farmers, which is why identifying the most appropriate tools for each situation is essential for its successful application.

Precision agriculture was the starting point for technological adoption at SLC Agrícola from 2003, with the adoption of grid soil sampling and the application of limestone at a variable rate. Thus, it evolved over the following years by adopting the variable rate application of other nutrients such as phosphorus and potassium. In 2009, with the massive introduction of the satellite differential signal, the use of on-board computers, autopilot, and satellite signal receivers, there were considerable advances. Among the benefits achieved by the company are the rational use of inputs, greater efficiency in operations, and reduced production costs.

In 2010, we produced the first cotton yield map in Brazil, obtained through a flow sensor installed in a harvester (Fig. 17.1). The use of these maps in the company enabled the identification of spatial variability and correlation with various production factors (Sana 2013). In 2015, inspired by the concepts of Industry 4.0, SLC began a series of tests of new technologies, which allowed it, in 2017, to implement various technologies for digital management. The objective was to meet the processes of agronomic management, machine management, phytosanitary management, and climate management, among others. Since then, the digital transformation of agricultural processes at SLC has had as its main challenges: encouraging technicians in the use of technologies, promoting the engagement of teams, and enabling the constant training of this team.

Fig. 17.1 Cotton yield maps, 2010/11 crop (**a**) and 2011/12 crop (**b**). (Source: Collection of the author of this chapter)

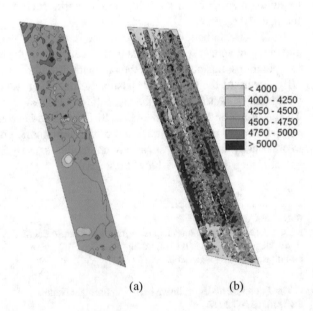

< 4000
4000 - 4250
4250 - 4500
4500 - 4750
4750 - 5000
> 5000

(a)   (b)

17.3 Planning for the Adoption of Digital Agriculture

The identification of the biggest problems to be solved in rural property makes it possible to select the most appropriate technologies that will deliver the best cost/benefit ratio in the short and medium term. From the selection of technologies to their implementation, SLC Agrícola uses the following steps to organize the process:

- Identify the issues to be resolved.
- Prioritize solving problems with the greatest impact on the business.
- Select company(ies) for testing.
- Carry out the test or proof of concept (PoC).
- Analyze technical and economic results.
- Decide on technological evolution and deployment plan.
- Control and measure benefits constantly.

Some tools can be used to support all steps. When identifying and prioritizing ideas and problems to be solved, the use of Design Thinking and prototyping techniques (Brown 2017) has proven to be efficient. In the Proof of Concept (PoC) planning and management stage, the Business Model Canvas (Osterwalder and Pigneur 2020) can be used. To improve and simplify the processes, the use of Lean Process techniques (Nielsen and Pejstrup 2018) is recommended.

17.4 Digital Implementation of Digital Agriculture Projects

The unrestricted support of the leaders or the owner and the directing of financial, personnel, and time allocation resources are fundamental for the implementation of Digital Agriculture projects. After the test results of the technologies have been satisfactory, implementation planning is necessary, which requires controls related to project management. Organizational change management is essential in the adoption of technologies, as it enables efficient communication with impacted technicians, proper training of teams and, therefore, greater engagement in the project (Miller 2012). The deployment phase of a technology is also known as Rollout, that is, the execution of the deployment plan. At this stage, monitoring the adoption of technology using indicators is essential, as it allows identifying and correcting problems and minimizing risks.

17.5 Results

Below are some of the main tools and technologies used at SLC Agrícola, as well as their technical and economic benefits, whether by reducing production costs or increasing crop productivity.

17.5.1   Production Management

The digitization of agriculture starts with the use of agricultural management systems. The first stage is Agricultural Planning, in which we define all the resources, inputs, labor, and machinery needed for the next season. At this point, it is essential to define the investment capacity and the available budget. During the production cycle, the use of software that help in the management of rural property and in important processes, such as inventory management, agronomic data, use of inputs, supplies, and production input, becomes essential to compose the production cost.

17.5.2   Precision Agriculture

The process of soil sampling, fertility map generation, and soil correction at a variable rate with lime, phosphorus, and potassium provided, in recent years, a 5% increase in the productivity of the corrected areas. In Fig. 17.2 the clay and phosphorus maps and the variable rate recommendation are presented, aiming at soil correction.

Yield maps are the base layer for identifying variability and supporting agronomic recommendations. Figure 17.3 shows a map with regions of very high productivity, more than 4000 kg ha^{-1}, and regions with low productivity, less than 2000 kg ha^{-1}, indicating the possibility of identifying and correcting the problem or adjusting management zones. At SLC Agrícola, productivity maps covering an area of approximately 200,000 hectares are generated, analyzed, and incorporated into the company's database each year.

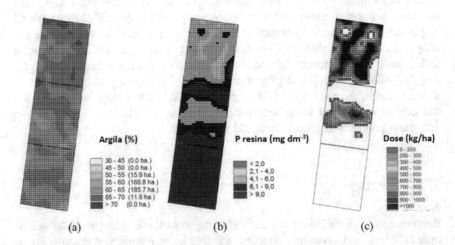

Fig. 17.2 Spatial variability maps of clay content (**a**), analysis of phosphorus (P) extracted from resin (**b**), and prescribed phosphorus dosage (**c**). (Source: SLC Agrícola)

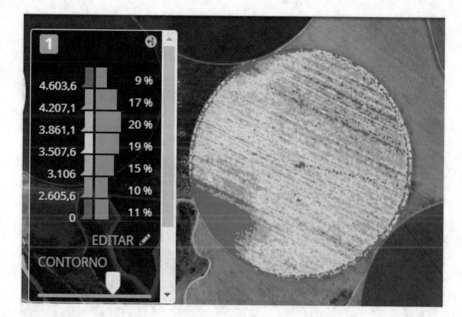

Fig. 17.3 Cotton productivity map (@/ha), in Cristalina, GO. (Source: SLC Agrícola)

The association of several data layers, such as clay content maps, productivity maps, and vegetation indices obtained in images, allows the definition of management zones and production environments, adapting the productive potential to the rational management of inputs.

17.5.3 Phytosanitary Management

The localized application of inputs is one of the most common practices in Digital Agriculture. The following are examples of opportunities for localized applications with agricultural pesticides that can be carried out during the season:

(a) *Selective Input Application Sensors*

The use of sensors installed on the sprayer bars that perform the application in real time allowed for a reduction between 80% and 95% of the volume of herbicide applied in pre-seeding application (Fig. 17.4). The operating principle is the emission of red and infrared lights by the sensor, which makes it possible to identify the chlorophyll in plants and, thus, carry out the targeted application of products. Sensors can be calibrated to run applications on the target crop and reduce the required volume of insecticides, growth regulators, and foliar fertilizers, especially when crops are in the vegetative phase.

(b) *Remote Sensing*

Fig. 17.4 Targeted and selective application of herbicide in the pre-sowing period. (Source: SLC Agrícola)

The use of satellite images, such as the images from the Sentinel-2 mission, makes it possible to monitor crops through maps of vegetation indices, such as the Normalized Difference Vegetation Index (NDVI) and the Soil Adjusted Vegetation Index (SAVI). With these indices, it is possible to identify variability in crops, which, whenever possible, should be associated with field observations, for the correct diagnosis or correlation with problems encountered during the harvest (Fig. 17.5). Among its main uses is the planning of precision agriculture for the following crops.

Another practical use of image-generated vegetation indices is selective and localized applications, which can be carried out during the crop cycle, such as top dressing, foliar fertilization, application of growth regulator, ripening, and defoliating. In Fig. 17.6, a cotton crop in the final stage of cultivation can be seen, in which the regions of the map with the highest vegetation index indicate the presence of leaves and, consequently, the need for defoliant application. In the area with the lowest vegetation index, the cotton is completely defoliated, eliminating the need for defoliants. This selective and differentiated application allowed reductions of 30–50% of the volume used in crops, with significant savings and reduction of environmental contamination. The prescription of localized application, using the vegetation indices, is carried out with Digital Agriculture software, in which it is possible to compare the quality of the operation carried out in the field (Fig. 17.7).

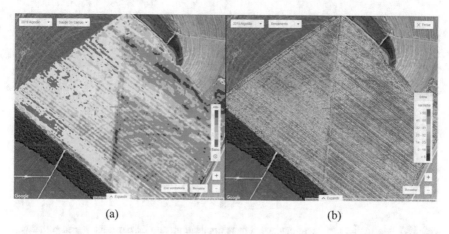

(a)   (b)

Fig. 17.5 Vegetation index map (**a**) compared to the cotton crop productivity map (**b**), Porto dos Gaúchos, MT. (Source: SLC Agrícola)

Fig. 17.6 Normalized Difference Vegetation Index Map (NDVI), Costa Rica, MS. (Source: SLC Agrícola)

Drones or Unmanned Aerial Vehicles (UAVs) have several uses in agriculture. Drones equipped with cameras with visible wavelength (RGB – Red, Green, Blue) allow simple, fast, and very useful inspections of crops, such as identifying failures in sowing or input applications. Drones equipped with multispectral cameras (RGB + Near Infrared) are capable of providing phytosanitary diagnosis of crops through vegetation indices such as NDVI (Fig. 17.8). There are also equipment that have high flight performance and can be used for planialtimetric planning and crop monitoring.

Fig. 17.7 On the left, the map planned for the application, and on the right, the map that was executed by the machine. Costa Rica, MS. (Source: SLC Agrícola)

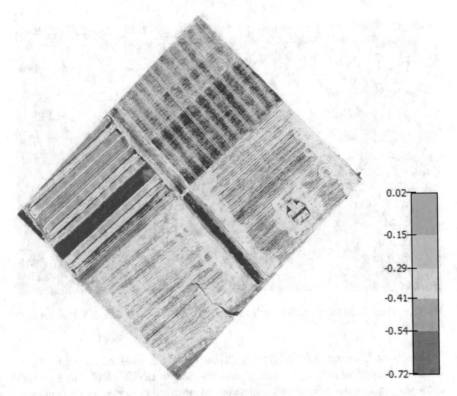

Fig. 17.8 NDVI maps, in cotton, obtained with a MAPIR® camera, in Diamantino, MT. (Source: SLC Agrícola)

(c) *Georeferenced Notes*

All observations recorded in crops, also called notes, must be carried out in a georeferenced way, as they allow visualization in heat maps, which can result in the selective and localized application of pesticides. The process is characterized by field monitoring by trained technicians to identify pests, diseases, weeds, nutritional deficiencies, or other plant stresses. These technicians go through the areas identifying these stresses and quantifying them, as well as noting their georeferencing. At the end of the process, the data are synchronized on the cloud platform and enable the generation of heat maps, which in turn can be exported in a compatible format for reading on agricultural machines, aiming at the rational application of pesticides (Fig. 17.9). The greatest savings opportunities in this area are in the localized application of insecticides, which, in the case of SLC Agrícola, has already been able to reduce up to 20% the volume of the product applied in soybean crops.

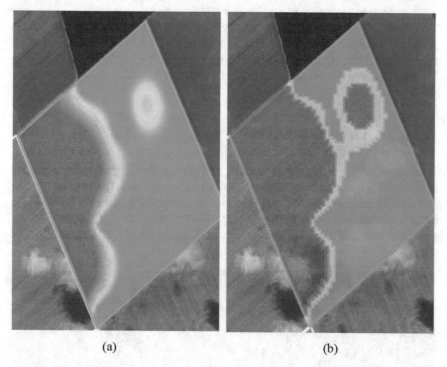

(a)   (b)

Fig. 17.9 Heat map and monitoring points (**a**) and prescription map for localized application (**b**). (Source: SLC Agrícola)

17.5.4 Machine Management

Currently, agricultural machines are equipped with several sensors for monitoring their components and for operation. This means that it is possible to monitor the performance of the machine, the operator, and the operation being carried out. In Fig. 17.10, a soybean sowing map can be seen, with variation in the rate of seeds deposited in each seeder section.

Operational errors, such as speeding, failures in seed placement, and lack of regulation of the machines, can be analyzed on Digital Agriculture platforms. Depending on the availability of rural connectivity, it is possible to obtain the information in real time (Fig. 17.11) or extract the data from the machines and import it into the agricultural management platforms.

In aerial spraying, the traceability of agricultural input applications is essential to validate the quality of the operation. In Fig. 17.12, an application can be seen that presented adequate quality indices, with 97.71% accuracy.

Still on the issue of monitoring agricultural machines, sensors can inform premature wear of parts, as well as the approaching date of periodic equipment maintenance, such as changing belts, changing oil, or cleaning filters, thus avoiding the need to stop the equipment in critical moments of its use.

17.5.5 Climate Management

Weather stations, soil moisture sensors, soil probes, and digital rain gauges are being used for climate management by agricultural companies. The weather forecast with greater assertiveness and real-time precipitation recording makes it possible to make important decisions, such as the application of inputs at the right time and when and where to carry out sowing. For example, in Fig. 17.13 a meteorological

Fig. 17.10 Soybean sowing map in Correntina, BA. (Source: SLC Agrícola)

Fig. 17.11 Machine and operations management platform, in Diamantino, MT. (Source: SLC Agrícola)

Fig. 17.12 Application map and indicators of aerial spraying, in Chapadão do Sul, MS. (Source: SLC Agrícola)

station installed in a cultivation area is shown transmitting real-time data and a climate management panel.

Stations onboard sprayers are decisive to indicate whether the application conditions of crop protection products are within the recommended parameters, based on information such as temperature, humidity, wind speed, and direction through the Delta T – a reliable and efficient indicator of the amount of steam that the atmosphere can absorb at a certain temperature. From an agronomic point of view, this information is essential to minimize possible pesticide drift.

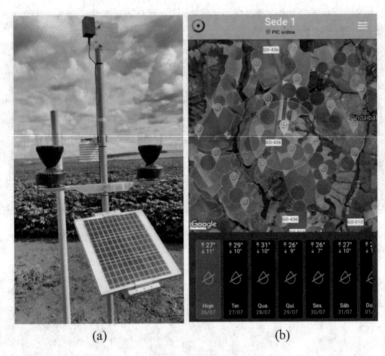

Fig. 17.13 Meteorological station (**a**) and data management platform (**b**). (Source: SLC Agrícola)

17.6 Indicator Management

The large amount of data generated by sensors, machines, notes, and images must be organized, integrated, and made available in visible management indicators. The indicators of Digital Agriculture platforms can help in the following decisions:

- Operational – Compilation of data or alerts for decision-making in a short period of time and, thus, the correction of problems (Fig. 17.14).
- Tactics – Indicators with cost/benefit information for management.
- Strategic – Indicate trends in agricultural and financial planning for future crops.

In large cotton, soybean, and corn production operations, it is necessary to monitor indicators through the Agricultural Operations Centers (COA) or Agricultural Intelligence Centers (CIA). In these places, the indicators are presented on panels and screens, enabling the agronomist to monitor the entire process and interact with the field teams, aiming to maintain the indicators and targets at excellent levels.

Fig. 17.14 Operational management indicator showing the variation in the dose of agricultural inputs. (Source: SLC Agrícola)

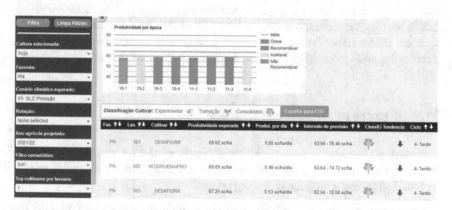

Fig. 17.15 Decision-making tool based on artificial intelligence. (Source: SLC Agrícola)

17.7 From Data to Prediction

Agriculture is undergoing a major digital transformation, in which the use of technologies such as Big Data (large amount of data), cloud computing, artificial intelligence, and machine learning enables the analysis and statistical modeling of large amounts, allowing decision-making to be performed with more accuracy and precision. The next step is the prediction of decision scenarios, based on historical data and forecast information. According to Diamandis (2019), all technologies, over time, become faster and cheaper and have greater computational power.

A practical example is the AIA tool, developed by SLC Agrícola (Fig. 17.15), with the support of a startup. The history of research and production of crops, data on varieties and cultivars, yield and quality of production, and analyses of soil fertility and soil nematodes and rainfall were used. The database was organized and, using data science techniques, algorithms were created capable of recommending the best variety for each agricultural company and field, based on the history and forecast of climatic macro-phenomena, such as el Niño and la Niña.

17.8 People Management

The true digital transformation is the transformation of people, as they are the ones who use technology and improve processes. However, when introducing a new technology, a key factor is to employ change management techniques, working toward reducing the time and impact of adoption. Training, communication, and engagement actions with Digital Agriculture tools are decisive for success and ensure people's digital inclusion.

Another skill demanded by managers is to remove traditional barriers, such as those indicated in the expressions: "I've always done it this way," "this is something from another sector," etc., and promote a collaborative environment between teams. In this journey, new professionals become part of the daily routine of agriculture, such as Information Technology professionals, data scientists and specialists, technicians, and agronomists prepared to interact with technologies.

17.9 Future of Digital Agriculture

Increased productivity in current cultivation areas, combined with the rational use of inputs, will make agriculture more efficient and competitive. This evolution will be achieved by improving the following points:

- Rural connectivity – High potential to increase Digital Agriculture, making the cycle between data collection and decision-making fast and reliable. The Internet of Things (IoT) protocols or even 5G networks will enable communication between machines, sensors, and systems instantly.
- Robotics – Autonomous machines performing operations in the field and using algorithms for real-time decisions look like movie scenes, but they are already real in agriculture (The Economist 2016). The introduction of these machines on a large scale will depend on reducing the cost of equipment, adapting the technology to the legal conditions of each country, and trusting in information security since data and equipment can be accessed remotely.
- Innovation – Innovating requires effort, learning, and tolerance for mistakes. Farmers will be able to use technologies from startups and companies, but they will also be able to use their knowledge and that of their team, to develop ideas and improvements, which have a low cost but great impact on the business.
- Traceability – Consumers are demanding information related to the origin and quality of products, in addition to the credibility of farmers and agricultural

companies. Digital Agriculture provides gains in agility and assertiveness in the production chain.

- Sustainability – Seeking efficiency in agricultural production, reducing production costs and increasing productivity, is to be competitive in agribusiness. The company, by being more efficient, reduces its environmental impact, is more sustainable, and contributes to guaranteeing future generations the capacity to meet the production needs and quality of life on the planet.

Abbreviations/Definitions
- Drones/UAVs: An *Unmanned Aerial Vehicle* (UAV) is an aerial vehicle that does not carry a human operator.
- NDVI: Normalized Difference Vegetation Index, which is a dimensionless index that describes the difference between visible and near-infrared reflectance of vegetation cover and can be used to estimate the density of green on an area of land.
- RGB: The spectrum of the visible light.

Take Home Message/Key Points
- SLC Agricola, a farming group, has experienced digital transformation and found it to be useful in improving crop management.
- Digital farming improved machinery and human resources management.
- Although some of the older staff showed difficulty in accepting the changes it proved to be easier for their tasks after the learning phase.

References

Brown T (2017) Design thinking: a powerful methodology for decreeing the end of old ideas. Alta Books, 272 p

Diamondis P (2019) The power & implications of exponential change. Disponível em: https://www.diamandis.com/blog/are-you-an-exponential-entrepreneur

Miller D (2012) Successful change management. Integrare, 208 p

Nielsen VF, Pejstrup S (2018) Lean in agriculture: create more value with less work on the farm. Productivity Press, 180 p

Osterwalder A, Pigneur Y (2020) Business model generation: innovation in business models. Alta Books, São Paulo, 300 p

Sana RS (2013) Special variability of soil and crop attributes on cotton yield using precision farming tools in the Brazilian cerrado. Dissertation (Masters in Soil Science) – Universidade Federal do Rio Grande do Sul, Porto Alegre

The Economist (2016) Technology quarterly: future of agriculture. Disponível em https://www.economist.com/technology-quarterly/2016-06-09/factory-fresh

Index

© The Editor(s) (if applicable) and The Author(s), under exclusive license to
Springer Nature Switzerland AG 2022
D. Marçal de Queiroz et al. (eds.), *Digital Farming*,
https://doi.org/10.1007/978-3-031-14533-9

Printed in the United States
by Baker & Taylor Publisher Services